GOTTLOB FREGE

The Arguments of the Philosophers

EDITOR: TED HONDERICH

The purpose of this series is to provide a contemporary assessment and history of the entire course of philosophical thought. Each book constitutes a detailed, critical introduction to the work of a philosopher of major influence and significance.

Plato	J. C. B. Gosling
Augustine	Christopher Kirwan
The Presocratic Philosophers	Jonathan Barnes
Plotinus	Lloyd P. Gerson
The Sceptics	R. J. Hankinson
Socrates	Gerasimos Xenophon Santas
Berkeley	George Pitcher
Descartes	Margaret Dauler Wilson
Hobbes	Tom Sorell
Locke	Michael Ayers
Spinoza	R. J. Delahunty
Bentham	Ross Harrison
Hume	Barry Stroud
Butler	Terence Penelhum
John Stuart Mill	John Skorupski
Thomas Reid	Keith Lehrer
Kant	Ralph C. S. Walker
Hegel	M. J. Inwood
Schopenhauer	D. W. Hamlyn
Kierkegaard	Alastair Hannay
Nietzsche	Richard Schacht
Karl Marx	Allen W. Wood
Gottlob Frege	Hans D. Sluga
Meinong	Reinhardt Grossmann
Husserl	David Bell
G. E. Moore	Thomas Baldwin
Wittgenstein	Robert J. Fogelin
Russell	Mark Sainsbury
William James	Graham Bird
Peirce	Christopher Hookway
Santayana	Timothy L. S. Sprigge
Dewey	J. E. Tiles
Bergson	A. R. Lacey
J. L. Austin	G. J. Warnock
Karl Popper	Anthony O'Hear
Ayer	John Foster
Sartre	Peter Caws

GOTTLOB FREGE

The Arguments of the Philosophers

Hans D. Sluga

London and New York

First published 1980 by Routledge

This edition reprinted in hardback 1999, 2001
by Routledge
2 Park Square, Milton Park, Abingdon, Oxon, OX14 4RN

Simultaneously published in the USA and Canada
by Routledge
270 Madison Ave, New York NY 10016

Transferred to Digital Printing 2009

Routledge is an imprint of the Taylor & Francis Group

© 1980 Hans D. Sluga

All rights reserved. No part of this book may be reprinted or reproduced or utilized in any form or by any electronic, mechanical, or other means, now known or hereafter invented, including photocopying and recording, or in any information storage or retrieval system, without permission in writing from the publishers.

British Library Cataloguing in Publication Data
A catalogue record for this book is available from the British Library

Library of Congress Cataloguing in Publication Data
A catalogue record for this book has been requested

ISBN 0-415-20392-9 (set)

ISBN10: 0-415-20374-0 (hbk)
ISBN10: 0-415-48772-2 (pbk)

ISBN13: 978-0-415-20374-6 (hbk)
ISBN13: 978-0-415-48772-6 (pbk)

Publisher's Note
The publisher has gone to great lengths to ensure the quality of this reprint but points out that some imperfections in the original may be apparent.

To ask historically means this: to release the happening resting and bound in the question and to set it in motion.

Martin Heidegger

Contents

	Preface	*page*	ix
	Introduction		1
1	*Analytic philosophy: a family tree*		1
2	*The non-historical character of analytic philosophy*		2
3	*How analytic philosophy misinterprets its own history*		4
4	*Two philosophical conceptions of language*		6
I	Philosophy in Question		8
1	*The three phases of nineteenth-century philosophy*		8
2	*Logic in German philosophy*		10
3	*The triumph of idealism*		12
4	*Philosophy in transition*		15
5	*The emergence of naturalism*		17
6	*Gruppe's linguistic naturalism*		19
7	*Czolbe's physiological psychologism*		26
8	*The decline of naturalism*		32
II	Philosophical Reconstruction		35
1	*New movements of late-nineteenth-century philosophy*		35
2	*Frege's foundational interests*		41
3	*Frege's pre-Begriffsschrift writings*		44
4	*Trendelenburg and the idea of a logical language*		48
5	*Lotze's conception of a pure logic*		52
6	*The influence of Leibniz and Kant on Frege's thought*		58
7	*Logical formalism as the root of the analytic tradition*		61
III	A Language of Pure Thought		65
1	*Frege's idea of a* Begriffsschrift		65
2	*Logical symbolism and language*		71
3	*Frege's propositional logic*		76
4	*The rejection of subject and predicate*		82

CONTENTS

	5 *Frege's logical notion of a function*	85
	6 *The analysis of general sentences*	87
	7 *The priority of judgments over concepts*	90
IV	In Search of Logical Objects	96
	1 *Frege's methodological critique of mathematics*	96
	2 *The strategy of Frege's anti-empiricism*	100
	3 *Logical laws as universal*	108
	4 *Logical laws as laws of truth*	112
	5 *Frege's Lotzean notion of objectivity*	117
	6 *Frege's concept of an object*	121
	7 *The logical analysis of natural numbers*	123
V	The Analysis of Meaning	129
	1 *The development of Frege's views in the 1890s*	129
	2 *The relation between a theory and its constitutive concepts*	130
	3 *The assignment of logical structure*	134
	4 *Concepts as functions*	137
	5 *Functions and value-ranges*	144
	6 *Value-ranges and set theory*	146
	7 *The logical justification of value-ranges*	149
	8 *Truth as the reference of sentences*	157
VI	The End of Logicism	162
	1 *Russell's contradiction*	162
	2 *Frege's way out*	165
	3 *The search for new foundations*	170
	4 *Russell*	176
	5 *Carnap*	178
	6 *Wittgenstein*	181
	Notes	187
	Index	199

Preface

This book is the result of a long development. For almost twenty years my interest in Frege has been part of a journey from Munich through Oxford and London to Berkeley. At each stage my thoughts about Frege were influenced by teachers, colleagues, and students. When it comes to acknowledging their help, I feel at a loss to describe it adequately.

My interest in Frege was aroused by one of my first philosophy teachers, Wilhelm Britzlmayr of the University of Munich. A close friend of Heinrich Scholz—the collector of Frege's writings when they were largely ignored—Britzlmayr had gathered a circle of students to work on Frege's logic and philosophy of mathematics. In this circle I first learnt to admire Frege's clarity of thought and precision of style.

At Oxford, in Michael Dummett's seminar, I became acquainted with a more probing, critical examination of Frege's thought. The continuing philosophical debate drew my attention for the first time to Frege's theory of meaning. For some time my reading of Frege was strongly colored by Dummett's tastes and outlook. If I find myself now in disagreement with some of his claims, I must still acknowledge that he has taught me an independence of mind that I value.

When I went to University College London, my friends Hide Ishiguro, Jerry Cohen, and Richard Wollheim provided new stimuli by making me think about Frege in relation to Leibnizian rationalism, nineteenth-century materialism, and late-nineteenth-century idealism. As editor of the series in which this book is now appearing, Ted Honderich first encouraged me to undertake the project of writing it and then nursed it along with infinite patience. Of particular importance for me was the contact with Imre Lakatos and Paul Feyerabend. Their critical assessment of the prevailing philosophy of science and their insistence that science could be understood only as a historical phenomenon raised in my mind the question whether a similar criticism could not be made of the account analytic philosophers had given of philosophical theorizing.

PREFACE

These reflections were of the greatest importance in the formation of the outlook adopted in this book.

In Berkeley this direction of thought was reinforced by Barry Stroud, who encouraged me in the study of Wittgenstein and helped me to understand his ideas more deeply. He and other colleagues and friends, in particular Hubert Dreyfus and John Searle, prodded me on with encouragement and critical questions. Berkeley in the early 1970s was a place of intellectual turmoil. I believe that being there forced me to think more radically about Frege and the analytic tradition that has come out of his work. The demands of my students for clarification and elaboration proved immensely helpful in the process. Among those students Peter Eggenberger perhaps most deserves notice for his unfailing interest and his many helpful suggestions.

As I look over the final result it seems to me that I can recognize in it also the subtle influence of Oskar Becker, my very first teacher in philosophy. I remember with admiration his profound scholarship in logic and mathematics, combined with a deep appreciation of Heidegger's historical understanding of the human world.

By recording these various influences I am trying to make clear that my book has a wider focus than is usual in the literature on Frege. Given the role Fregean considerations have played and continue to play in logic, the philosophy of mathematics, and the philosophy of language, it is difficult to write comprehensively about Frege's thought. In the pursuit of such a comprehensive picture I have tried to take into account as much of the existing literature as possible, but I have not tried to detail and discuss all the important issues raised in it. Throughout the book I have found it necessary to refer back to Michael Dummett's writings on Frege, both because they reflect an important stage in the current discussion of the topic and also because I consider the shortcomings of his account symptomatic. However, my discussion does not presuppose detailed knowledge of his extensive and difficult writings.

One chief difference in my treatment of Frege as compared with Dummett's and that of most of the literature is its greater historical orientation. I hope to show that this approach generates a sharper and more coherent picture of Frege's thought, a picture that differs from the predominant one in substance and not only in minor additions and revisions. I also hope that this reassessment will be of more than historical interest. It seems to me plausible to assume that we always understand ourselves only historically, i.e., as creatures who have reached the present moment via a certain route. If one could show that we have reached that moment along a path different from the one we believe we have traveled, that would deeply change our understanding of who we are. My purpose in trying to reassess the roots of the analytic tradition is to change the current self-interpretation of that tradition.

PREFACE

One difficulty in writing about Frege from a historical point of view is the vast literature that needs to be surveyed for that purpose. The writings of nineteenth-century philosophers are extensive, difficult and, for us, curiously opaque. Men like Trendelenburg and Lotze who influenced Frege's ideas are hardly known today, though they were once famous. Others like Gruppe, Czolbe, or Mauthner are almost completely forgotten. So far there exists no adequate account of the movements of thought that influenced Frege or to which he reacted. What follows will therefore certainly be in need of augmentation and revision.

My study is incomplete also in another respect. Its aim is to describe a development that begins with Frege, but I have only a few sketchy things to say about the later stages of that development. My remarks about Russell, Carnap, and Wittgenstein are loose and unsystematic. In my mind I see a line of descent that leads from Frege to Wittgenstein and finds there both culmination and termination. The few references to Wittgenstein give no adequate indication of this picture. They also fail to show the extent to which my procedures have benefited from Wittgenstein's thought. It must be left to another occasion to provide the missing details.

Apart from the philosophical debt I owe to Barry Stroud, I am also obliged to him for many stylistic suggestions that have helped to make my formulations more precise and the text as a whole more readable.

In discussing Frege's views I have largely adopted the English terminology chosen by Black and Geach—not because it is inevitably the best, but because it is so entrenched that a change would only create new confusions. At the same time I have generally preferred my own translations of the Fregean texts I quote. In fact, apart from following the Kemp-Smith translation of the *Critique of Pure Reason*, I have translated almost all German texts quoted, even where English translations are available. My references are, therefore, consistently to the German originals rather than to their translations. In referring to Frege's writings I have found it convenient to identify texts consistently by reference to the standard German editions published by Olms and Meiner.

The following abbreviations have been used:

BS   *Begriffsschrift*, ed. I. Angelelli, Hildesheim, 1964.
F   *Die Grundlagen der Arithmetik* [The foundations of arithmetic], ed. and trans. by J. L. Austin, Oxford, 1959.
GG *Die Grundgesetze der Arithmetik*, vols 1 and 2, Hildesheim, 1962.
KS *Kleine Schriften*, ed. I. Angelelli, Hildesheim, 1967.
NS *Nachgelassene Schriften*, ed. H. Hermes et al., Hamburg, 1969.
WB *Wissenschaftlicher Briefwechsel*, ed. G. Gabriel et al., Hamburg, 1976.

Introduction

1 Analytic philosophy: a family tree

First, philosophers thought about the world. Next, they reflected on the way the world is recognized. Finally, they turned their attention to the medium in which such recognition is expressed. There seems to be a natural progression in philosophy from metaphysics, through epistemology, to the philosophy of language.

Whether we consider this development necessary and beneficial for philosophy or not, the fact remains that philosophers since the beginning of the nineteenth century have increasingly turned their attention to language. In this process metaphysical and epistemological doctrines have often been reformulated in terms of the philosophy of language.

Philosophical concern with language has progressed on a broad front. For Herder it was associated with an interest in the history of mankind, for Gruppe with the advances of philology and the critique of Hegelian metaphysics. Frege thought about language in conjunction with his foundational investigations in logic and mathematics. At the same time Mauthner's reflections on language were motivated by Mach's critical epistemology. In the twentieth century philosophers like Carnap and Tarski concerned themselves with the study of the syntactic and semantic properties of formal languages, Wittgenstein thought about the language of everyday life, and Heidegger about the language of the poet. In the last thirty years philosophers have produced a rich harvest of theories about language which range from theories that offer recursive descriptions of the syntax and semantics of formal and natural languages, through theories that deal with language in terms of truth and verification, or intention and speech act, or in causal and referential terms, to theories that set out to investigate the deep structure of actual discourse.

Of all the groups of philosophers who have been concerned with language the largest and most coherent is that of analytic philosophy. It is

INTRODUCTION

the characteristic tenet of that school that the philosophy of language is the foundation of all the rest of philosophy. Analytic philosophy has tended to concern itself with language in a particular manner, holding that the structural, formal, logical investigation of language is philosophically fundamental. For analytic philosophers there has tended to be a close link between the philosophy of language and the study of logic and the foundations of mathematics. The question in what follows is to understand how analytic philosophers came to focus on this particular aspect of language.

2 The non-historical character of analytic philosophy

Gottlob Frege can be considered the first analytic philosopher. His thoughts about language had a profound influence on such diverse thinkers as Wittgenstein and Carnap, who might well be considered the opposed poles of the analytic tradition. His influence can also be felt in the logical and mathematical writings of Russell and, more recently, in the efforts of analytic philosophers to construct a comprehensive theory of meaning. Through his critique of Husserl's early psychologism he also contributed to the formation of phenomenological thought.

Granted that Frege is a seminal figure in analytic philosophy and that he has a significant place in both the history of logic and the history of philosophy of the last hundred years, we may ask: why has it taken so long for this fact to be recognized? Why is it even now not fully acknowledged?

Among the reasons for this curious neglect is the analytic tradition's lack of interest in historical questions—even in the question of its own roots. Anti-historicism has been part of the baggage of the tradition since Frege. In spite of a large body of analytic literature on the topic of Frege—including several recent books—we are still far away from a detailed historical picture of the sources of Fregean ideas, of the overall structure of his thought, and of the contribution it has made to the subsequent development of logic and philosophy. The reason is that the literature has tended to discuss Frege's writings predominantly in the context of the contemporary problem situation. It asks what contribution Frege's text makes to this or that contemporary issue. This may be helpful if our concern is to advance the philosophy of language, but what if we wish to understand the historical roots and the dynamics of the philosophical movement that has come out of Frege's thought?

There are deep reasons why the writings of the analytic tradition are unhelpful at this point. From its very beginning, the tradition has been oriented towards an abstract, formal account of language and meaning, and not towards the comprehending of concrete historical processes. Frege himself considered his task that of the analysis of timeless, objective thoughts. 'The thought is something impersonal,' he wrote in 1897. 'When we see the sentence "2 + 3 = 5" written on a wall, we recognize the thought expressed by it completely and it is altogether irrelevant for our understanding to know who wrote it' (NS, p. 146). If we apply this doctrine to the analysis of statements in philosophical texts, we are led to conclude that the thoughts expressed by them are objective, timeless, and impersonal, however subjective, historical, and personal their formulation may be. The task of the interpreter is then seen as freeing the thought from its subjective, historical clothing and thereby laying it open to assessment. The method of interpretation will become that of rational reconstruction.

Michael Dummett's extensive discussion of Frege and the philosophy of language can serve as a paradigm for the failure of analytic philosophers to come to grips with the actual, historical Frege.[1] No one has discussed Frege's thought at greater length; there is much to be learnt about Frege in his writings; he pursues Fregean problems with a great deal of philosophical acumen; and yet, in the end, he fails to give a compelling historical picture. What he says turns out to be motivated more by his own search for an anti-realist theory of meaning than by the wish to uncover the real historical significance of Frege's writings.

There is one notable step Dummett has taken beyond Fregean objectivism. Influenced by the later Wittgenstein and by the Dutch intuitionist Brouwer, he argues that a theory of meaning cannot be a theory of objective, timeless thoughts, but must be a theory of understanding and that such understanding occurs and changes in time. One might therefore expect him to conclude that the investigation of the meaning of Frege's statements cannot consist in a rational reconstruction of the timeless, impersonal thoughts supposedly expressed by them, but that it must proceed on the assumption that the meaning of actual philosophical discourse is both historical and personal. Such expectations are disappointed, because Dummett's recognition of the temporality of thinking is limited by the formalist interpretation of that idea he has inherited from Kant via Brouwer. Temporality is understood in this Kantian tradition as abstract and formal, not as concrete historical time.

At least since Tarski's essay 'On the Concept of Truth in Formalized Languages' of 1931 it has become the common belief of linguistic philosophers that a purely formal, structural theory of meaning could be constructed. The meaning of every statement, it is said, is recursively determined by the meanings of its simple constituents. This belief has replaced the doubts of Frege and the early Wittgenstein about the possibility of a precise semantic theory. Whereas they treated semantic utterances as mere hints or as stepping-stones that must eventually be discarded, linguistic philosophy now holds that a rigorous semantic theory should be possible.

As long as one remains with a few simple, formalizable sentences the idea seems to hold promise. Analytic philosophers speak readily about meaning and interpretation; but their examples are carefully chosen pieces of discourse. One might, of course, argue that this procedure is designed to proceed systematically from simple to more complex cases. Eventually, it will therefore reach the analysis of the most difficult philosophical texts. But there is little evidence in the actual progress of the discussion to encourage such hopes. After eighty years of debate linguistic philosophers are still not agreed on the semantics of simple proper names. How long would it take to construct a semantics of philosophical discourse?

The conclusion seems, therefore, that the analytic techniques do not lend themselves to talk about the meaning and interpretation of philosophical texts—including those texts that make up the history of the analytic tradition itself. While this conclusion does not undermine actual insights of linguistic philosophy, it seems to subvert the *interpretation* of those insights which analytic philosophers have adopted. For it denies the possibility of the analytic theory of meaning being truly fundamental. A contemporary analytic philosopher reaches his problems through a training that involves interpreting texts which constitute the analytic tradition. The meaning of the contemporary problems is therefore a function of the meaning of the historical discourse within the tradition. But if that meaning is inaccessible with the tools of the analytic theory of meaning, then there is at least one sense in which that theory cannot be basic philosophy.

We can now understand why Dummett's recognition of the temporality of understanding does not lead him on to the full realization of the historical character of the meaning of philosophical discourse. For to admit that historical character would be to question the foundational nature of the analytic inquiry which Dummett correctly perceives as the precondition of the functioning of the analytic tradition. To question the foundational nature of a formal theory of meaning is to question the picture of philosophy as progressing from metaphysics, through epistemology, to the philosophy of language. We would then need to re-examine the role of the philosophy of language and that of philosophy in general, and to question the idea that from now on the examination of the logical structure of language will be the proper task of philosophy. The question why and how philosophers came to focus their attention on language may be a first step towards that re-examination.

3 How analytic philosophy misinterprets its own history

While Frege may be called the first linguistic philosopher, he was certainly not the first philosopher of language. His specific concern with language was stimulated by the historical context in which language had become of fundamental philosophical significance. Frege's particular kind of interest in language was interconnected with a number of other philosophical concerns, directly of an epistemological and indirectly of a metaphysical kind, that involve him with the philosophical problems of the late nineteenth century.

The historical circumstances of his thought separate him from the heyday of analytic philosophy. His ideas are deeply indebted to the thought of Hermann Lotze and through him to the philosophical constructions of Leibniz and Kant. In its *later* development the analytic tradition incorporated strongly empiricist elements into its thinking. Such empiricism was quite alien to Frege, but that of course is not to deny the great influence of Frege's thought on the formation and course of analytic philosophy.

A proper historical analysis will have to take account of Frege's mediating role between late-nineteenth-century philosophy and the linguistic philosophy of the present century. Because of their lack of historical interest, analytic philosophers themselves have tended to overestimate the discontinuity of their own philosophizing from that of the past and to underestimate the historical evolution of their own tradition. They have tended to characterize the analytic tradition as systematic and scientific in contrast to philosophical pre-history, which cannot legitimately lay claim to either of these adjectives, and to interpret the internal development of the tradition in terms of the logic of its own theorizing rather than in terms of historical causation.

In this sense, Moritz Schlick has spoken of the rise of analytic philosophy, with the definite article, as 'the turning point in philosophy.'[2] Using the same definite article, Ayer has spoken of it as 'the revolution in philosophy.'[3] And, in exactly this spirit, Dummett has analysed Frege's contribution to philosophy. He calls him, in a sense 'the first modern philosopher';[4] and elsewhere he adds that 'we can date . . . a whole epoch of philosophy beginning with the work of Frege, just as we can do with Descartes.'[5] Given the belief in a sharp separation between the analytic tradition and the rest of the philosophical past, it is not surprising that any attempt to relate Frege's thought to the issues of the nineteenth century will be seen as intended merely 'to serve the purpose of playing down his originality and placing him as one among many members of the school of "classical German philosophy."'[6] For instance, any suggestion of a link between Frege and Lotze must be rejected as a 'remarkable piece of misapplied history' (*ibid.*, p. 457). Instead, Dummett says, we must assume Frege's logic 'to have been born from Frege's brain unfertilized by external influences.'[7] Such judgments are unfortunately by no means idiosyncratic to Dummett; they seem to express a pervasive attitude of analytic philosophers towards the history of their tradition.

INTRODUCTION

The complementary tendency, that of underestimating the distance that separates the later tradition from its beginnings, can equally be illustrated in the case of Frege. Its effect is also that of blocking real historical understanding. Thus, it is taken for granted that Frege was concerned with ontological questions just as the subsequent analytic tradition has been.[8] It is assumed that he was interested in setting up a semantic theory just as logicians have done since Tarski, that, indeed, model-theoretical semantics begins with Frege.[9] His considerations about truth as an object are dismissed as mere scholasticism (*ibid.*, pp. 644-5). His rejection of logicism after the discovery of Russell's paradoxes is considered an overreaction; his objections to Cantorian sets are explained as the result of personal hostility. Wherever Frege's views can be made to fit the current discussion, they are simply identified with it; where they cannot be made to fit, they are either ignored or explained away in psychological terms.

The present study of Frege's thought, the investigation of why and how he came to think about language, is meant as more than an act of historical justice. By trying to show the continuities that tie the analytic tradition to the historical past as well as the discontinuities that separate it from philosophical pre-history, by setting out to show both the continuities and discontinuities within the tradition, it may be possible to reveal the *historical* character of the tradition itself and thereby to change the perception analytic philosophers have of their own tradition. Ultimate and fundamental as its concerns may look from a certain perspective, they have this character only in the given historical context. It is from this context that the analytic concerns have originated; it is in this context that they flourish; and with it they will eventually fade away.

4 *Two philosophical conceptions of language*

One of the first things we discover when we look at the historical roots of the analytic tradition is the fact that its concern with language is the outcome of two initially independent and, in certain ways, opposed developments. Both have their origin in the nineteenth century.

On the one hand, a philosophical interest in natural language developed at that time; on the other hand, an interest in precise, formal systems of notation, or logical languages, also flourished. While the former was motivated by historical and empirical concerns, the latter had its roots in mathematical and formal considerations. The interest in natural language was often associated with empiricist and naturalistic tendencies, whereas the interest in logical languages was often connected with anti-empiricist and rationalistic viewpoints. The particular way in which a philosopher talked about language was therefore connected with and depended on his metaphysical and epistemological convictions.

INTRODUCTION

The analytic concern with language has undergone several transformations since the beginning of this century. At first analytic philosophers focused almost entirely on formal languages. Later on, for a while, it was claimed that natural language is the philosophically interesting phenomenon and that talking about it in terms of formal languages was unhelpful and even misleading. Even more recently the focus has still been on natural language, but with the conviction that it possesses a deep structure which is regular and which can be brought out by comparing ordinary language to artificial formal languages.

At the same time the accompanying metaphysical and epistemological views have shifted a number of times. In Frege's thought the interest in logic is accompanied by a strong apriorism; later on the analytic tradition adopted more empiricist views and began to reinterpret *a priori* truths as empty tautologies or as true by convention; still later the distinction between *a priori* and *a posteriori* truths began to be questioned altogether. Metaphysically, viewpoints have shifted between a critical philosophy in the Kantian sense, empiricist reductionism and a naive and dogmatic ontologizing.

My concern in what follows is to illuminate some of these struggles and shifts by focusing on the Fregean contribution to the formation of the analytic tradition.

I

Philosophy in Question

1 The three phases of nineteenth-century philosophy

Today we can see that a powerful philosophical tradition has one of its roots in the work of Frege. This was not something that Frege himself ever came to realize. When he died in 1925 he was a bitter and disillusioned man. There are various reasons for his disappointment: some personal, some professional, and some historical.

In certain respects his fate was similar to that of two other philosopher–mathematicians, Cantor and Dedekind. At about the same time, in the last quarter of the nineteenth century, all three of them set out to provide deeper foundations for the theory of natural and real numbers. All three were motivated by an opposition to the prevailing philosophical climate, all three hoped to change it through their foundational investigations. And all three discovered the strength of the opposition to them. Frege's fate will concern us further. Cantor's was to be driven mad by the relentless attacks of his enemy Kronecker with his robust mathematical naturalism; Dedekind's to be buried alive in an insignificant post at the Technical College at Braunschweig.

What was it that made the reception of their ideas so difficult? With respect to Frege it has been hypothesized by Michael Dummett that it was the dominance of Hegelian idealism that obstructed the recognition of his achievements. Dummett writes: 'In a history of philosophy Frege would have to be classified as a member of the realist revolt against Hegelian idealism, a revolt which occurred some three decades earlier in Germany than in Britain.'[1] The suggestion is that Frege reacted against a dominant German Hegelianism in the 1870s in the same way as Moore and Russell reacted against British Hegelianism around 1900. In addition Dummett has argued that the psychologism which Frege attacks so frequently and with such vehemence was connected with idealist philosophy.[2]

> Idealism is by its very nature prone to slip into psychologism, although the possibility of a viable idealistic theory of meaning depends precisely upon the possibility of resisting this temptation. But in any case, in Frege's day the kind of idealism that was everywhere prevalent in the philosophical schools was infected with psychologism through and through: it was not until it had been decisively overthrown that it became possible to envisage a non-psychologistic version of idealism. . . . Hence it was almost certainly a historical necessity that the revolution which made the theory of meaning the foundation of philosophy should be accomplished by someone like Frege who had for idealism not an iota of sympathy.

These are the final words of Dummett's Frege book and therefore are presumably to be taken seriously as comments on the historical and intellectual position of Frege as a philosopher. Closer inspection of the facts, however, reveals that Dummett's hypothesis does not reflect the actual situation; for it is a mistake to think that Hegelianism (or any other form of idealism) was dominant in the period of Frege's activity and it is another mistake to consider psychologism connected with such a dominant idealism.[3]

Idealism had in fact ceased to be a real power in German thought by about 1830. What terminated at that time was not just a particular set of philosophical ideas, but a tradition that began with the inception of modern German philosophy by Leibniz. The development was one result of a deep shift in the pattern of European culture in the half-century between 1775 and 1825. Nowhere was the shift so close to the surface and therefore so noticeable as in Germany. The accelerated growth of the natural sciences, the rapid appearance of new technologies, the increase in population size with its accompanying changes in social and political relationships, urbanization, industrialization, nationalism and democracy, the loosening of the hold of Christian belief, the new sciences of man which seemed to account for human life in purely naturalistic terms — all of these reflected immediately on the standing of philosophy and put it into question.[4]

The question these developments posed was not just which of various possible philosophies was the most satisfactory, but whether there was any need, any place for philosophy at all. Ever since 1830 the position of philosophy in the culture has been at best insecure, at times marginal, sometimes openly challenged. Before that, sceptics had occasionally questioned the possibility of philosophy together with the possibility of any knowledge whatsoever, and believers had questioned its necessity in the face of divine revelation. But there had been agreement that, if there was a need for human knowledge and the possibility of it, philosophy had an important place in this undertaking. It was precisely this belief that the developments of the early nineteenth century undermined. Since then philosophers have continued to struggle with the question how philosophy is possible and why it is still necessary in the scientific–technological world.

In the years between 1830 and 1870 philosophy was wholly on the defensive in German thought. It was only after 1870 that philosophers found some security again. One way in which they tried to establish the usefulness of philosophy was by arguing that its task was the investigation of the logical structure of mathematics, science, and language. Philosophy was possible in the contemporary world as formal logic.

2 Logic in German philosophy

Leibniz's thought is commonly interpreted within the context of the Cartesian reconstruction of philosophy, but such an understanding overshadows some of the peculiar features of the Leibnizian synthesis which were to determine the later development of German philosophy. The philosophical climate in which Leibniz was raised was heavily impregnated with a conservative Aristotelianism. From this early influence Leibniz took into his own mature philosophizing an unwavering faith in the power of Aristotelian logic. Unlike Bacon or Descartes he never disparaged the value of logic, but, on the contrary, assigned to it a singularly important role in his own methodology. Against Descartes he wrote:[5]

> It seems to me fair that we should give to the ancients each one his due; not by a silence, malignant and injurious to ourselves, conceal their merits. Those things which Aristotle prescribed in his *Logic*, although not sufficient for discovery, are nevertheless almost sufficient for judging; at least, where he treats of necessary consequences; and it is important that the conclusions of the human mind be established as if by certain mathematical rules. And it has been noted by me that those who admit false reasonings in serious things more often sin in logical form than is commonly believed. Thus in order to avoid all errors there is need of nothing else than to use the most common rules of logic with great constancy and severity.

These strictures were directed, in the first instance, against Descartes's intuitionistic theory of truth, according to which a proposition is true if it is clear and distinct. This criterion, Leibniz claimed, is wholly subjective. 'I do not see what use it is to consider doubtful things as false. This would not be to cast aside prejudices, but to change them' (*ibid.*, p. 48). In place of Descartes's criterion an objective — and that means a logical — criterion must be proposed.

As to what is said by Descartes, that we must doubt all things in which there is the least uncertainty, it would be preferable to express it by this better and more expressive precept: We ought to think what degree of acceptance or dissent everything merits; or more simply: We ought to inquire after the reasons of any dogma (*ibid.*, p. 47).

Any truth must have a reason why it is true. For Leibniz that reason is to be found in the principle that in a true proposition the predicate is always contained in the subject.

Through Christian Wolff the Leibnizian conception of the role of logic was handed on to the later generations of German philosophy. It was through his mediation that Leibniz's three logical principles (identity, non-contradiction, and sufficient reason) became part of the stock-in-trade of German thought. The continuing preoccupation with logic in German philosophy of the eighteenth and nineteenth centuries is one feature that distinguishes that tradition very clearly from the French and British traditions and reinforced its rationalistic tendencies. It reveals itself not only in the fact that 'Transcendental Logic' occupies the central place in Kant's *Critique of Pure Reason*, but also in the speculative constructions of the German idealists for whom considerations about the principle of identity (Fichte, Schelling) and the principle of non-contradiction (Hegel) were fundamental. For Hegel logic was 'the pure science,'[6] whose concern was 'pure knowledge' or 'pure thought.' As objective logic it constituted the beginning of all science (*ibid.*, p. 59), taking the place formerly held by metaphysics (*ibid.*, p. 54).[7]

While the interest of philosophers surrounded logic with a great deal of speculation, it did very little to further the science itself. In the period between Leibniz and Frege no attempt was made at a fundamental reform of logic. There are several interlocking reasons for this failure. To begin with, Leibniz himself had never overcome the limitations of Aristotelian logic and the philosophical tradition that followed did not do any better. In his *Critique* Kant could still hold that since Aristotle logic had not been forced to retrace a single step: 'it is remarkable also that to the present day this logic has not been able to advance a single step, and is thus to all appearance a closed and completed body of doctrine' (B viii). The view expressed here was so powerful that it was only in the late nineteenth century, when the continuity of the philosophical tradition had been disrupted, that Frege could feel free to reconstruct his logic unhampered by the model of the syllogistic theory.

Since Leibniz had never been able to construct a completely satisfactory logical calculus, his technical writings remained uncompleted and, for the most part, unpublished. They saw the light only when the revival of logic in the nineteenth century drew attention to Leibniz's technical efforts in the subject. What had been known to the philosophical tradition were Leibniz's speculative uses of logic. Since his time philosophers had been concerned with the epistemological, metaphysical, and psychological implications of logic. The tendency had become so strong by the time of Kant that he felt it necessary to warn: 'We do not enlarge but disfigure sciences, if we allow them to trespass upon one another's territory. The sphere of logic is quite precisely delimited; its sole concern is to give an exhaustive exposition and a strict proof of the formal rules of all thought' (B x–xi).

Kant's insistence that logic is merely a formal science and must be distinguished sharply from non-formal, empirical sciences was to prove of great significance for the later development of the subject. It is no accident that Kant in fact coined the term 'formal logic' which is often used to contrast the kind of logic with which contemporary logicians are concerned with other types of logic (such as dialectical logic). In spite of Kant's call for a pure, formal logic, logic books of the nineteenth century tended to interpret the science in a very loose way. For instance, in Wundt's *Logik*—which had several editions between 1890 and 1920—the traditional theory of concept, judgment, and inference is still embedded in a wide-ranging discussion of notions such as representation, apperception, association of ideas, fact, substance, causality. It is only with the arrival of symbolic logic that the connection of the science with such issues is broken.

3 The triumph of idealism

Where Leibniz had been a universalist in philosophy, Kant thought of himself as a critical thinker. For Leibniz logic had been the key to the systematic and rational development of the totality of human knowledge. The difference between conceptual, necessary truths and factual truths had been only one of relative clarity and distinctness. Kant, under the influence of British empiricism, came to consider such claims unrealistic and postulated a sharp distinction between conceptual reasoning and sensory perception. Kant's critical philosophy assigned to itself a carefully delimited role. The legitimate uses of philosophical reasoning were to be accounted for in terms of the fundamental dichotomies of sensibility and understanding, analytic and synthetic, appearance and thing-in-itself.

The impact of Kant on the philosophical generation that followed him was profound, but ambiguous. The German idealists, foremost among them Fichte, Schelling, and Hegel, all recognized the Kantian origin of their philosophizing, but all of them came to reject the Kantian dichotomies. Where Kant had been concerned with separating philosophical inquiry from empirical knowledge, the German idealists set out once again to establish the unity of knowledge. Abandoning Kant's caution, they tried to show that all knowledge, whether abstract or empirical, could in the last resort be derived from the transcendental principles of consciousness. All knowledge was in the end *a priori*, necessary, philosophical. By arguing in this way, the idealists proved themselves closer to the spirit of Leibniz than to that of Kant. As a consequence their notion of logic tended to be wider than the narrow Kantian notion of formal logic, or even Kant's transcendental logic. Logic incorporated the totality of knowledge; reasoning was the only reliable method to attain it.

These views had quite extreme consequences. Soon Schelling was proving that there were exactly three forces in nature, magnetism, electricity, and galvanism, and that there were, by necessity, three dimensions in space, three periods of revelation, and so on. And he held that these facts could be derived from the principle of the identity of the self, an instance of the logical principle of identity.[8] Such efforts were soon followed by the *Naturphilosophen* (themselves often respectable scientists such as the anatomist Lorenz Oken), who criticized in the severest terms the empirical method of Newton and endeavored to reconstruct the natural sciences in a purely systematic manner. Louis Agassiz, the biologist, described the effect of Oken's lectures on his audience:[9]

> Among the most fascinating of our professors was Oken. A master in art of teaching, he exercised an almost irresistible influence over his students. Constructing the universe out of his brain, deducing from *a priori* conceptions all the relations of the three kingdoms into which he divided all living beings . . . it seemed to us who listened that the slow laborious process of accumulating precise detailed knowledge could only be the work of drones, while a generous, commanding spirit might build the world out of its own powerful imagination.

These ideas culminated, of course, in the system of Hegel, who cast no less a spell over his audiences. Not only the natural world, but the historical world and the social and intellectual history of mankind were brought by him into a systematic deductive pattern. For a whole generation the prospects of this kind of universalistic rationalism seemed irresistible.

Eventually, however, the development reached the point of crisis. In 1831, at the height of his fame, Hegel was struck down by cholera. But even without his sudden death change would have come. There were (1) political, (2) social, (3) scientific, and (4) internal philosophical reasons for it. The program embodied in the idealist tradition which had once looked promising had now begun to deteriorate drastically. (1) In the thinking of the time, Hegel's system had become aligned with the authoritarian regime of Prussia. After the July revolution of 1830 had shaken the repressive governments established by the Vienna Congress, political criticism began to be heard all over Germany. Out of it came the political attacks made on the Hegelian system by Heine, Feuerbach, and later Marx and Engels. (2) Under cover of the post-Vienna Congress regimes Germany had been changing from the country of the romantic poets to a pre-industrialized society. Technological and scientific progress were in the air. When the Prussian king called the aging Schelling to Berlin in 1841 to combat the pernicious intellectual developments of the time, the once-acclaimed philosopher found almost no audience and quickly faded from the scene. Idealism seemed no longer suited to the spirit of the times. (3) Even in 1827 Alexander von Humboldt had attacked speculative *Naturphilosophie* before overflow audiences at the University of Berlin, where he showed how the deductive conclusions of Hegel and Schelling conflicted directly with the observational results of the growing sciences. Scientists like Johannes Müller and Justus von Liebig were actively engaged in promoting new observational and experimental techniques, rejecting the deductive methods of the idealists.[10] (4) In the face of such counter-evidence, the derivations of the idealists began to look philosophically weak and vague and to amount to mere empty verbiage. In addition, it seemed that those who were trying to adhere to the Hegelian and idealist tradition were unable to produce new variations or elaborations. Their writings appeared as eclectic and repetitive restatements of doctrines already formulated by the first generation of idealists.

The result was threefold: first, a turning away from idealist philosophy and from Hegelianism in particular; second, the rejection of the speculative, deductive, *a priori* method the idealists had used; and third, in so far as philosophy as a whole was identified with idealism and with deductive *a priori* reasoning, a turning away from philosophy as a whole. In the thinking of the times, idealism was replaced by materialism, *a priori* reasoning by empiricism, and philosophy as a separate intellectual activity by an ideology in which philosophy had merged with and disappeared in the empirical sciences. The ideology that replaced idealism could be called *scientific naturalism*. This ideology presented itself as a new and better kind of philosophy, and at other times as a scientific world view transcending any kind of philosophy. In retrospect we have come to see that it was after all no more than another philosophy and that, despite its apparently decisive victory, it had serious problems of its own which eventually led to its decline.

The generation of philosophers that grew up after the middle of the century was to see those weaknesses more clearly than the first generation of naturalists. By 1870 the reaction had set in. Frege's thought was conceived in opposition to this form of scientific naturalism, and not to a dominant Hegelianism or idealism, as Dummett has claimed.

In opposing themselves to scientific naturalism the philosophers of the late nineteenth century were often in sympathy with some doctrines of the idealists. That is why idealist and rationalist elements can be found in Frege's writings. The reaction against naturalism was not confined to Germany. It also gained ascendence in England, where, in philosophy, it stimulated a somewhat surprising interest in Hegel.[11] Moore and Russell were originally under the spell of British Hegelianism and revolted against it around the turn of the century. They rejected Bradley's monistic idealism and turned (at least initially) to an extreme form of pluralistic realism. Because of the connections between Frege and Russell it is easy to hypothesize that both were reacting against Hegelianism and that both set against it a philosophical realism. But Frege and Russell do not belong to the same phase in the development of the analytic tradition. And Frege was concerned neither with the formulation of an anti-idealist philosophy nor with a defence of realism.

4 Philosophy in transition

The collapse of idealism, the subsequent dominance of naturalism, and the slow recovery of philosophy can be illustrated most convincingly through statements of contemporaries. Justus von Liebig (1803–73), chemist:[12]

> I, too, spent part of my years of study at a University where the greatest metaphysician of the century [i.e., Schelling] inspired academic youth to enthusiasm and imitation. Who could resist this contagion? So rich in words and ideas and so poor as real knowledge and serious study were concerned. It cost me two precious years of my life. I cannot describe the shock and consternation upon awakening from this delirium. How many of the most gifted youths have I not seen perish in this fraud, and how many laments over lives completely ruined did I not hear later!

Rudolf Haym (1820–1901), historian of philosophy:[13]

> The truth is that the realm of philosophy is in a state of *complete anarchy*, in a state of dissolution and chaotic decay.

Matthias Schleiden (1804–81), biologist, anti-materialist:[14]

> I have no hope for the near future. The principled and to a certain degree justified aversion of our scientific–industrial age against all philosophical investigations cannot so easily be overcome.

Jacob Moleschott (1822–93), physiologist, materialist:[15]

> Philosophers can no longer be viewed in opposition to natural scientists, because any philosophy worthy of its name will lap up the best sap of the tree of knowledge, and thereby produce only the ripest fruit of that tree.

Wilhelm Wundt (1832–1920), philosopher, psychologist:[16]

During my time as a student and for a long time afterwards during my time as lecturer there existed only one professor of philosophy [at Heidelberg], who therefore represented the field in its entirety. It was Alexander Baron von Reichlin-Meldegg. He had once been a Catholic theologian who had converted to protestantism. . . . Meldegg preferred to lecture on systematic topics, whereby he combined several into one course, whenever possible. One of them, e.g., had the title: Psychology under inclusion of somatology and the theory of the mental disorders. Despite the superficiality of his lectures he never lacked students, since many of them were required to register in four philosophy courses during their studies. Of course, the students for the most part only registered for the lectures without attending them. . . . In spite of this he was occasionally commissioned by the Senate of the university to deliver a commemorative address in the auditorium because of his qualification as the only official philosopher. I still remember his speech on May 19, 1862, on the occasion of Fichte's one-hundredth birthday which, with its profuse exclamations addressing themselves to the shades of Fichte, probably was the most miserable of all the miserable speeches given that day.

After Reichlin-Meldegg's departure Eduard Zeller arrived as his successor from Marburg [1862]. A feeling of satisfaction began to spread among those interested in philosophy. With Zeller a man had been appointed who at that time was regarded as the pre-eminent representative of his subject in Germany. However, it cannot be said that he completely satisfied the expectations he had aroused. His manner of delivery was dry and lacked any kind of stimulating effect on the audience, because he dictated his lectures from beginning to end. . . . In his main course he lectured on the history of philosophy from Thales to Hegel in a single semester. . . . Zeller's own philosophical viewpoint . . . corresponded most closely to the views of Christian Wolff if one imagines his popular rationalism translated from the eighteenth to the nineteenth century and decorated with a little Hegelian dialectic.

When, after Zeller's departure [1872], Kuno Fischer was called from Jena, the position of philosophy in the university underwent a complete transformation. In him a man had been appointed who could captivate the academic youth with his brilliant teaching and speaking talent and who, with his gift for the written and even more the spoken word, could create a newly awakening interest in philosophy in wider circles. Kuno Fischer was himself no outstanding philosopher, but he was an incomparable interpreter of the philosophers, particularly of German idealist philosophy from Kant onwards. . . . From Reichlin-

Meldegg's course on Faust to Kuno Fischer's lectures on the same topic, a more extreme change can hardly be imagined. And yet it reflected the changing situation of philosophy itself. From its decline after the death of Hegel, through an interim period of respectful toleration, it has become the one subject of instruction in the university that can claim to have the most general educational value.

5 The emergence of naturalism

The anti-idealist revolt had begun in Germany with David Friedrich Strauss's *Life of Jesus Critically Examined* (1835–6) and was soon philosophically reinforced by Feuerbach's *Essence of Christianity* (1841). For Feuerbach the Hegelian system is the culmination of rationalism and, like all its forms, is basically inspired by a religious spirit. But God is only human nature projected into objectivity; the mind only a property of a concrete body, the idea only the creation of an actual human being. 'I do not generate the object from the thought, but the thought from the object; and I hold that alone to be an object which has an existence beyond one's own brain.'[17] Philosophy must cease to be speculative and become anthropological. Man is a sensory being; his knowledge comes from sensory perception of reality and not from speculative constructions out of concepts.

> Religion [and all speculative thought] is the dream of the human mind. But even while dreaming we are not in heaven or in the realm of Nothingness. We are right here on earth, in the realm of reality, even if we see real things not as they really are and as they must necessarily be, but in the enrapturing light of wishful imagination. Hence, I do nothing more to religion than open its eyes, or, rather, to direct its inwardly turned gaze toward what is outside, so that the object as it exists in the imagination changes into what this object is in reality (*ibid.*).

Philosophical naturalism thus originated in the critique of the Hegelian system as the paradigm of metaphysical speculation. This attack on the false consciousness of speculative thought eventually issued in the materialism of Marx and Engels. Meanwhile, by 1848 the development had already reached a second phase. The critique of Hegel had done its job. A second generation of writers, all of whom considered themselves indebted to Feuerbach, began to develop a naturalism based on the results of the empirical sciences rather than on the exposition of the internal difficulties of Hegelian metaphysics. Karl Vogt, Jakob Moleschott, Ludwig Büchner, and Heinrich Czolbe were the leaders of this movement.[18] All of them lived into the 1890s and pursued their ideas in a large number of popular scientific and philosophical writings.

Vogt, Moleschott, Büchner, and Czolbe were all physiologists by training, and that background clearly influenced their philosophical viewpoint. The development of physics, chemistry, and physiology in Germany by the middle of the nineteenth century in fact greatly influenced the general course of philosophical thinking. Liebig's work in organic chemistry seemed to support the belief that organic phenomena were merely physical and chemical processes of a highly complex form, though Liebig himself stuck to a vitalist conception. The advances in physiology made by Johannes Müller and his associates seemed to confirm the thesis that organisms are complex physiological mechanisms. Robert Mayer's discovery of the principle of the conservation of energy was interpreted as a proof of the impossibility of dualism, for the interaction of body and mind would involve energy loss or accretion in the physical system.

In England scientific naturalism drew much of its strength from Darwin's *Origin of Species*. But when that work appeared in 1859 there was already well established in Germany a scientific naturalism based on the structural and functional characteristics of the human organism, rather than on genetic, evolutionary considerations. On the whole, the German naturalists found confirmation in the teachings of Darwin, though some, like Czolbe, rejected the idea of evolution altogether and others, like Büchner, reinterpreted it in Lamarckian terms.

Philosophically, the naturalists concluded that their epistemology had to be a radical empiricism. They agreed with Feuerbach that human knowledge is built entirely on a sensory foundation. Consequently, they felt sympathetic to the British empiricists, in particular Locke and Hume. At the same time they were committing themselves to a strict ontological realism and materialism. Sensations, they argued, are just another kind of material phenomenon. Their view of human mental activity was quite definite. Thought was the natural product of the activity of the brain, just as urine was the natural product of the kidneys.

With such considerations in mind they hoped to be able to show the absurdity of traditional philosophy with its *a priori* reasonings and its belief in the fundamental character of logic. Concepts, they held, were just reflections of what was seen, of sensory activity. They were nothing without external referents. There were, of course, in their eyes no innate ideas. Even mathematical concepts had to be considered as rooted in experience. The laws of thought were identical with the mechanical laws of external nature. Logical laws were no more than empirical generalizations concerning human mental activity and that activity in turn was to be interpreted in physiological concepts. Psychologism (which Dummett associates with German idealism) was in fact a direct product of the naturalism of the middle of the century.[19] The same is true of mathematical formalism, Frege's other great target of attack. If the formulas of mathematics could not reasonably be interpreted as expressing psychological regularities, they had to be taken as concrete, uninterpreted symbols.

There was no room left for *a priori* truths and *a priori* concepts. What the individual might regard as *a priori* knowledge was nothing but the experience of the species, as the naturalists repeated after Spencer. Human reasoning was all based on induction. It was induction that led from concrete facts and observations to natural laws. It was induction that led from accidental truths to the so-called necessary truths of reason.

When Kant came back into fashion after 1870 the scientific naturalists had no sympathy for Kantian apriorism. 'This so-called "Back to Kant" is surely the saddest *testimonium paupertatis* that modern philosophy could give itself,' Büchner wrote.[20] Czolbe claimed to recognize a theological bent dominating Kant's logic, just as Feuerbach had criticized Hegel's idealism as theological. Neo-Kantianism was seen as a retreat into idealism and rationalism. Their own sympathies were rather with the radical empiricism of Mill. Mill's *Logic* of 1843 has been highly recommended by Liebig for its inductive methodology. Helmholtz had referred to it in his physiological optics and several translations of the work had been published in German.[21]

There is no need at this point for an exhaustive account of the doctrines of scientific naturalism.[22] Two topics, however, demand our closer attention. One is the development of the philosophy of language in the nineteenth century, the other the issue of psychologism. In what follows they will be discussed by referring to two representative figures, the philosophers Otto Friedrich Gruppe and Heinrich Czolbe. In Gruppe the critique of Hegel and the adoption of a naturalistic viewpoint is combined with a profound interest in the role of language in human understanding. For that reason Gruppe must be considered a philosopher of language; his views anticipate in various respects those of Mauthner and the later Wittgenstein. At the same time Gruppe's conception of language is in stark contrast to Frege's. By considering them together we can see the poles between which philosophy of language moves in the nineteenth century. In Czolbe's writings there is hardly any emphasis on language, but he is of interest because of his systematic naturalistic theory of knowledge. Czolbe's thought can be considered a classical expression of psychologism. It demands our attention specifically because of the critical attention devoted to it by Lotze with arguments that are directly reflected in Frege's writings.

6 *Gruppe's linguistic naturalism*

Given that philosophical practice is linguistic, that philosophers are continually engaged in speaking, writing, and reading, and that philosophical thinking essentially involves an operating with verbal signs, one wonders why philosophical reflection about language is such a recent phenomenon.

It is true that philosophers like Plato, Aristotle, Leibniz, and Locke reflected on language, but none of them could be called a linguistic philosopher in the modern sense, nor did their precedent initiate a continuous philosophical concern with language. Where one might have expected philosophers of the past to talk about language, they usually spoke of impressions, ideas, concepts, thoughts, propositions, judgments — all invisible items for which language was said to provide merely the outer clothing.

In Germany it was probably Herder who first saw deeper than this and succeeded in establishing a tradition which might be called philosophy of language. For Herder history was the key to understanding of the human world. Human beings are historical beings and this is revealed in the fact that historically grown languages are the immediate expression of the human mind. 'The human soul thinks with words,' he wrote in 1799. 'It does not just express itself, but identifies itself and orders its own thoughts by means of language. Leibniz says that language is the mirror of human understanding, and one may boldly add that it is also the source of its concepts — not only the customary but the indispensable tool of its reason. By means of language have we learnt to think, and through it we separate and connect concepts.'[23] Metaphysics therefore becomes the philosophy of human language (*ibid.*, p. 184). The origins of language are to be found not in a timeless reality but in time, and language forever carries the signs of this origin with it. 'Is it possible to ignore these traces of the changeable language-creating mind and to look for a ground in the clouds? What proof is there of a single word that God could have invented? Is there even one language in which there is even one single pure general concept which has come to human beings from heaven?'[24]

With the reflections of Herder (and the contemporary thoughts of Wilhelm von Humboldt) there begins in Germany a continuous development which eventually leads to the thought of Frege and Wittgenstein. It would hardly be correct to regard the philosophy of language as the dominant aspect of German philosophy in the nineteenth century, but it remained throughout the century as a continuous thread of thought.

Herder himself had set out to interpret the cultural peculiarities of different nations in terms of their distinct languages. These efforts had spawned a wealth of philological studies: the philology of the ancient languages by Wolff and Lachmann, that of Sanskrit by Schlegel and Bopp, and finally the philological study of German itself under the leadership of the brothers Grimm. Such studies were always motivated by philosophical concerns and sometimes gave rise to systematic reflections on the role of language in human life. The poet Novalis, the anarchist

Max Stirner, but most notably the now almost-forgotten Otto Friedrich Gruppe saw in language the direct expression of human consciousness.

Gruppe was born in Danzig in 1804 and died in Berlin in 1876. As a student he attended Hegel's lectures, but his main interest lay in the study of history and classical and German philology. He seems to have been particularly close to Karl Lachmann and early on came under the influence of Alexander von Humboldt's anti-idealism. These interests were combined in his first philosophical work, published in 1831, which contains what is perhaps the most sustained attack on the Hegelian system ever written. This early critique of Hegelianism may have cost him his academic career. Its title is *Antäus. Ein Briefwechsel über spekulative Philosophie in ihrem Konflikt mit Wissenschaft und Sprache* [A correspondence on the conflict of speculative philosophy with science and language].[25] Three years later he followed with *Wendepunkt der Philosophie im neunzehnten Jahrhundert* [The turning-point of philosophy in the nineteenth century]. In 1855 he published his last philosophical book, *Gegenwart und Zukunft der Philosophie in Deutschland* [The present condition and future of philosophy in Germany]. He was also the author of a large number of (now-forgotten) historical, philological, and literary works.

The third of his philosophical books was written on the occasion of Schelling's death in 1854. The once-acclaimed philosopher had died in complete obscurity, having outlived his own fame. With respect to the state of philosophy Gruppe asks: 'How are things, how have they become what they are, and where do they point?' (*Gegenwart*, p. iv). There is, he diagnoses, not only contempt for idealism, but a general lack of interest in all kinds of philosophy. The situation needs to be taken seriously by anyone 'whom the hostile divisions and dominating extremes have left with any faith in the possibility of an intellectually free and nevertheless unpresumptuous kind of philosophy' (*ibid.*, pp. iv–v). Since the time of Leibniz Germany had been dominated by a succession of philosophical systems. In each period one of them seemed to rule in the German universities, influencing all academic subjects as well as literature and the general cultural life. This strange phenomenon had been considered abroad as almost a peculiarity of the German national character. But now the last system had lost its glamour, no new one had stepped into its place, and there was no expectation of a new commanding authority, a new philosophical system that might subjugate once again everything around it (*ibid.*, p. 2).

Gruppe concludes that the decline in the standing of philosophy is due not to this or that minor error in the previously popular philosophies, but to the basic project philosophers had set themselves. 'If one considers that systems of speculative philosophy have long disappeared in England and France, and lately also in Germany . . . the thought may well arise whether perhaps the age of the speculative systems is over' (*ibid.*, p. 43).

In the end he concludes that 'there cannot be any more speculative systems and we have no cause to regret that fact' (*ibid.*, p. 258). For the history of thought is not, as Hegel thought, one that proceeds according to an inner law and contains truth at every stage; 'on the contrary, it is a history of error with occasional glimpses of insight' (*ibid.*, p. 256). Twenty-one years earlier, when this development had been much less clear, he had already written: 'It is now two years since I declared war in my *Antäus* against the efforts of all those men, most of them German, who call themselves speculative philosophers and claim under this name nothing less than rule over all the sciences' (*Wendepunkt*, p. 1). And he had added: 'My opposition is directed against metaphysics in general and against speculative philosophy which believes that it can draw knowledge out of mere concepts' (*ibid.*, p. 12).

What is the cause of the idealistic confusions? Gruppe sees the error deeply embedded in the history of philosophy. It can be discerned already in Aristotelian metaphysics and, through the influence of Aristotle, it spread into later philosophy (*ibid.*, p. iv). The flaw lies in the logical doctrines that form the fundamental framework of Aristotelian thought. 'Aristotelian logic achieves so little that it misses the true nature of judgment and of progressive knowledge altogether' (*ibid.*, p. vii). If philosophy is to be freed from its disastrous errors, a radical reconstruction of logic is inevitable. 'The sickness of philosophy is due to a false method. . . . Logic, epistemology must lay the foundations; we can expect no good without their thorough reform' (*Gegenwart*, p. 179). He hopes that quietly 'some isolated researcher has already turned to the logical tasks and that soon a fresher wind will blow' (*ibid.*, pp. 182–3). Much needs to be thrown out of philosophy, but logic remains as a legitimate field of philosophical inquiry. 'There is no doubt that the foundations for a complete reconstruction of philosophy lie in that specialized area' (*ibid.*, pp. 263–4).

The reform of logic, Gruppe envisages, must arise out of the study of language. Natural science and the comparative study of language together will defeat the errors of speculative thought. 'Only these two can bring about the understanding of the basis and essence of those errors which can be found in Aristotle and even in his predecessors' (*Wendepunkt*, p. vi). The study of language is important because speculative metaphysics entirely misunderstands language and its role in thinking (*ibid.*, p. 18). Thinking is not possible without language and language is not possible without thinking (*ibid.*, p. 28). Language is not something invented, something fabricated, it is not given prior to thought, nor is it a ready-made tool and organ of thinking; the one has grown together with the other (*ibid.*, p. 72). Speculative philosophy regards thinking as the original act of an even more original substance, which is called soul; and it regards language as the servant, the mere tool of thinking, as a sensory medium. In order to explain what lies open to view (the word), speculative philosophy flees to something unknown (the thought, the meaning, the idea) and from that to something even more unknown (the soul, the mind) (*Antäus*, pp. 314–15).

> The speculative error—in other words, idealism—has its origin in the closing of eyes and ears. Then the self is isolated and one wants to construct knowledge, philosophical systems, even the world out of the self.... But the self in isolation is something entirely empty, a mere abstraction, nothing real: the thinking self has grown up in the contemplation of the surrounding things. The isolation which Descartes demands as the beginning of speculative thought is not so much dangerous as entirely impossible (*Gegenwart*, p. 239).

The separation of language and thinking leads to the view that there are ready-made meanings which are conveyed by means of linguistic representation. But what we mean is made definite only by the language in which that meaning exists. For that reason we should be suspicious of the assumption that language may be given a speculative use at all (*Antäus*, p. 339). It is perfectly all right to speak in everyday contexts of the truth of an assertion. 'But the philosophers fell over the word and ripped it out of its roots through which it was attached to the ground of language and from which it gains its living and healthy meaning' (*ibid.*, p. 352). They were then led to the bizarre conclusion that there is truth as such, truth independent of what we say and assert. In his *Antäus* Gruppe sets out to show how Hegelian philosophy separates concepts such as 'space,' 'time,' 'quantity,' and 'nothing' from their living context and is thereby led into erroneous speculative conclusions. The common, everyday use of the word 'nothing' is perfectly in order. 'But when logicians and metaphysicians take this linguistic form in and by itself outside its context, it is their own fault if they are led into temptation' (*ibid.*, p. 325).

The fundamental insight needed to overcome the errors of speculative thought is the discovery that judgments, not concepts, are primary and that concepts function and are used only in judgments. Gruppe writes:

> The basic evil lies in logic and in particular in a false theory of abstract concepts. Traditional logic consists essentially of three parts: namely, the doctrines of concept, of judgment, and of syllogism. It regards concepts as the simplest elements out of which judgments are composed, and out of those, in turn, syllogisms. The proper investigation must go quite a different route; for it the judgment is the simplest; concepts originate only in judgments and only there is the act of concept formation to be sought and found (*Gegenwart*, p. 191).

Since traditional logic conceives of judgments as combinations of concepts, it must regard concepts as ready-made before the judgment.

But 'concepts are merely the results of judgment and . . . can only be explained through them' (*ibid.*, p. 43). That insight should be the starting-point of a radically different kind of logic (*Wendepunkt*, p. 93).

Because concepts originate in judgments, 'classifications into kinds are not immediately discovered or given, they do not constitute divisions and partitions made by nature itself, but they express only our own viewpoint as the result of combination and comparison. In a word, they are nothing but our judgments' (*ibid.*, p. 61). Concepts do not lie in things themselves but are relations between things as they are grasped and judged by us (*ibid.*, p. 100). There are no innate and no *a priori* concepts, as Plato, Leibniz, and Kant had thought. 'The generality and necessity of concepts of which there is so much talk proves to be nothing but a misunderstanding of that relativity of concepts which hangs essentially together with the nature of judgment and the origin of language' (*ibid.*, p. 81). That does not mean that our use of concepts is altogether arbitrary. For example, they must come in pairs of opposites to be useful to us, however fleeting reality itself may be (*Antäus*, pp. 329–30). And they possess a *certain* generality and a *certain* necessity, but both the generality and the necessity of our concepts arise only out of our common human need, out of practical use (*ibid.*, pp. 348–9).

When we consider, on the basis of such assumptions, what concepts themselves are, we must say that they are nothing but mortgages on particular things (*ibid.*, p. 317). They are basically metaphors. When we judge a thing to be an *A* we mean to compare it to something else that we have previously called an *A*. Similarities and other relationships impose themselves on us naturally when we speak in this way. As the scope of our judgments grows our concepts come to cover new similarities and so continuously grow and expand.

> [Concepts] have begun from pictures and metaphors and even in their ripe old age they remain comparisons, even if in their condensed form they no longer point to sensory objects. But that does not affect their basically metaphorical nature. Therefore the same thing holds of them as of ordinary pictures: one must abandon them after they have done their work, one cannot employ them permanently, one must not expect any thoroughgoing consistency from them (*ibid.*, p. 365).

Speculative philosophy arises when language no longer fulfills its acceptable elementary function. In ordinary discourse our terms function well enough. They allow us to express ever-new recognitions of resemblances. In philosophy we reflect on our use of concepts. Speculative thought takes them out of the context in which language has introduced them. Then everything collapses. The history of speculative metaphysics is dominated by the idea that concepts are fixed universals and that therefore things either fall under the concept or do not. The belief in the unrestricted validity of the principle of excluded middle and the use of indirect argument based on that belief lie behind many of the characteristic teachings of traditional philosophy (*ibid.*, p. 344; *Gegenwart*, p. 226–7). Thus, Leibniz had argued that there must be simple things, since there are complex things. But

> One cannot infer simple objects from complex ones. One cannot say that there *are* simple things; one cannot even say in a metaphysical sense that there are complex *things*. Why? It is correct and innocent to call a house, a pillar, or a chemical substance either complex or simple. And there is no harm in calling houses, pillars, and chemical substances things. But the matter changes radically when (like Leibniz) we speak quite generally, without qualification, of complex things which exist (*Antäus*, p. 338).

And what is wrong with that? It is wrong because all our disjunctions of terms depend on this or that comparison, which is not given with the things themselves. Disjunctions depend always on a certain context. In no way is there *one* class of things which are complex and another class which are simple, as Leibniz had maintained (*Antäus*, pp. 337–8).

Gruppe's philosophy of language originated from the same assumptions as that of scientific naturalists like Vogt, Moleschott, Büchner, and Czolbe.[26] Nevertheless, his views differ in certain important respects from theirs. Gruppe's naturalism is not the outcome of an interest in physiology. Like Feuerbach he is motivated above all by the critique of Hegel. Unlike most other naturalists Gruppe has a genuine interest in logic. Given his call for the rejection of Aristotelian logic and for a radical reform of the subject, he can be said to have sketched a program that Frege was later to carry out. But Frege carried it out in a manner quite different from that envisaged by Gruppe. For Frege this reform involves the construction of a formal language of objective conceptual contents. His model for the reform of logic is the mathematical calculus. Behind that conception lies the belief that meanings are separable from the language in which they are expressed and that language can be judged by how well it expresses such meanings. Gruppe's conception of logic is more strictly linguistic. His conception of the reform of logic is based on a critical study of natural language. The failure of Aristotelian logic in his view is not that it falls short of some ideal standard but that it does not reflect the nature of reasoning as it proceeds in actual language.

In one important point Gruppe diverges from the predominant views of naturalism and empiricism. The naturalists generally agreed with the traditional empiricist doctrine that judgments are combinations of concepts and that, more generally speaking, human mental activity proceeds by composition from simple to complex ideas. Gruppe had been sufficiently influenced by Kant to find that position implausible. He holds with Kant that judgments possess an original unity and that concepts result from the analysis of judgments. Gruppe's conception was eventually taken up by Trendelenburg and actually played an important role in the anti-naturalist reaction of the late nineteenth century. In Gruppe's own thinking, however, the doctrine has a quite different function since it is conjoined with basically naturalistic assumptions.

This combination of naturalism with anti-atomism gives Gruppe's ideas a strikingly Wittgensteinian appearance. There are indeed a number of surprising similarities in their thought: (a) the critique of metaphysics as arising out of the misuse of language and the accompanying claim that ordinary language is philosophically acceptable; (b) the idea of a reconstruction of philosophy through the study of the logic of language; (c) the rejection of empiricist atomism and the claim that there is no sense in which one can speak of absolutely simple objects; (d) the explanation of concepts as expressing resemblances (a view shared by Nietzsche and Mauthner); (e) the account of the necessity of concepts as arising out of 'human needs'; (f) the critique of indirect argument as a characteristically metaphysical method of proof; (g) a theory of concepts according to which they always come in pairs of opposites. To point out such similarities is of course not to deny the differences between the two philosophers. They are separated by a century. Gruppe was trained as a philologist with a scientific and historical interest in language, and he clearly was a less original and powerful thinker than Wittgenstein.

There is no reason to think that Wittgenstein was at all familiar with Gruppe's writings, but he did know Mauthner's works, in which very similar ideas are to be found. Mauthner himself fully recognized the affinity with Gruppe's ideas, though, as he pointed out, his own work had not been influenced directly by Gruppe. In any case it remains true that besides the tradition of the philosophy of language which is most easily identified in Frege there was another tradition, represented most clearly by Gruppe and Mauthner, which in Wittgenstein became eventually intertwined with the Fregean strand.

7 *Czolbe's physiological psychologism*

When the naturalists read Mill they particularly felt the appeal of Mill's psychological interpretation of logic and mathematics. According to Mill's famous statement,[27] logic

> is not a science distinct from, and co-ordinate with Psychology. So far as it is a science at all, it is a part, or branch of Psychology; differing from it, on the one hand, as a part differs from the whole, and on the other, as an Art differs from science. Its theoretical grounds are wholly borrowed from Psychology, and include as much of that science as is required to justify the rules of that art.

The German naturalists could agree entirely with that statement, but for them it meant something quite different from what it meant for Mill. Mill's statement had been written against the background of the British empiricist tradition. For him psychology was concerned with interior, subjective phenomena, with sensations, sense-data, and ideas. The German naturalists also considered themselves committed to empiricism, but they were realists and materialists at the same time. Psychology, in their eyes, was concerned with objective, material phenomena which were ultimately to be interpreted in physiological terms. This becomes quite clear as soon as we turn to the writings of Czolbe.

Though both Gruppe and Czolbe can be considered as part of the naturalist tradition, their views differ in important respects. Czolbe is not much interested in the fate of traditional philosophy or in the critique of Hegel. And where Gruppe is motivated by the understanding of the logic of language, Czolbe is guided by the progress of physiology. There is, however, one thing that ties him to the tradition. Of the scientific naturalists he is the only one who felt the need to develop his philosophical convictions systematically.[28] It was not that the others were incapable of such an undertaking, but they were more concerned with the public dissemination of the insights of scientific materialism, with social and political reform, even with the improvement of health and nutrition. From Feuerbach to Büchner they all subscribed to Marx's famous saying that the philosophers had only tried to understand the world, but that the real task was to change it. Czolbe, politically and socially more conservative than the others, was given more to theoretical reflection. As a result he has given us the material through which we can understand the strengths and weaknesses of the naturalistic position.

Czolbe, who was born in 1819 and died in 1873, studied medicine and physiology and obtained his doctorate under the physiologist Johannes Müller. Throughout his medical career he retained a lively interest in philosophy and this interest convinced him that the progress of science required the construction of a 'system of naturalism.'[29] There existed, he argued, no single presentation of naturalism which 'determines precisely its fundamental principle and answers the most important questions about the connections between the issues, as is done in philosophical systems. What Feuerbach, Vogt, Moleschott and others have done in this respect most recently are only suggestive, fragmentary claims which leave one dissatisfied when one enters the subject more deeply' (*ibid.*, pp. v–vi). Strauss, Bruno Bauer, and Feuerbach had led him to conceive his naturalistic system. In addition there was the influence of the philosopher Lotze.[30] In his *Allgemeine Pathologie* of 1842 Lotze had attacked vitalism and argued that *all* natural processes could be explained in a completely mechanical manner. He thought there was no need to refer to a higher principle, a life force, in the *scientific* analysis of life processes. His book

PHILOSOPHY IN QUESTION

had gained the respect of contemporary physiologists.[31] It also impressed Czolbe and became the stimulus for his own thought. Generalizing Lotze's conclusions, he formulated the fundamental principle of his sensualism: *the total exclusion of the suprasensual* (*Neue Darstellung*, p. 203). He fully realized that this had not been the intended consequence of Lotze's argument, for in his *Medical Psychology* of 1852 Lotze had added to the views of his earlier book the thesis that mechanical explanations could never be ultimate and had to be supported by an idealist metaphysics. Lotze's position was therefore akin to that of Leibniz, who had allowed that science must be mechanical but also stands in need of metaphysical foundations. *Medical Psychology* had disappointed the physiologists of the modern school.[32] It had also motivated Czolbe to compose his *Neue Darstellung des Sensualismus* 'as a kind of positive refutation' (*Neue Darstellung*, p. viii).

The exclusion of the suprasensual which is the basic postulate of the work means the acceptance of the existence only of what is in principle observable (*ibid.*, p. 1). According to Czolbe the acceptance of the suprasensual in addition to intuitive (*anschaulich*) concepts is one of the main sources of false reasoning. It has led to extravagant superstitions; in fact, all the mysterious conceptions of the mind and of nature are based on this dualistic view. Its ultimate source is to be found in traditional logic which recognizes the existence of the suprasensual (*ibid.*, pp. 2, 62).

The traditional logic is to be opposed by the new sensualism. Its principle is, of course, not capable of exact proof. It can only be 'explained' and made plausible. It cannot be deductively justified (*ibid.*, p. 2).

> One might call it even a prejudice or a preconceived opinion. But without such a prejudice the formation of any view of the connections between the phenomena is impossible. When scientists believe that they form concepts, judgments, and inferences out of sensory perceptions without any preconceived opinion, this belief rests only on self-deception (*ibid.*, p. 6).

There are basically two such preconceptions: sensualism and suprasensualism. All human understanding presupposes the one or the other. 'It is impossible to conceive of a logic without one of these two principles' (*ibid.*). Because the sensualistic principle is not provable, suprasensualism always remains a possibility (*ibid.*, pp. 61–2). It remains attractive, because it provides one with easy explanations of the world by postulating unknown suprasensual quantities (*ibid.*, p. vii). But too much value is attached to the concept of possibility; it is the ever-open door leading to innumerable sophistries. The systems of religion and speculative philosophy are never logically impossible. That is why it is useless to try to refute them from the standpoint of traditional logic (*ibid.*, p. 233).

What is needed is to develop a plausible version of sensualism. The development of a coherent system of naturalism is the only way to defeat the suprasensualist systems of thought. Sensualism remains little more than a rather indifferent slogan for the deeper requirements of science as long as it does not show in detail *how* perceptions, ideas, concepts, judgments, inferences, the will, etc. arise through the senses (*ibid.*, pp. vi–vii).

With this argument Czolbe tries to justify the more systematic philosophical procedure he adopts in his book. It sets his thought apart both from the polemically and practically oriented writings of other naturalists of the period and from a simple empiricist faith in experience. Whether Czolbe's attempt to show *how* sensualism can be made to work was successful (and could have been so) is, of course, another question.

The fundamental given for Czolbe's sensualism is sensory perception (*sinnliche Wahrnehmung*). Concepts are for it only makeshift devices and they are best when they stay close to perception (*ibid.*, p. 3). For that reason *psychology is the fundamental science* (*ibid.*, p. 8), not logic, not speculative thought. With this stress on psychology and sensations and with Czolbe's affinity with radical empiricism one might easily take him to be a kind of phenomenalist. But sensations are not more fundamental to him than the material world, they are just part of it. Sensualism and materialism are, for Czolbe, identical (*ibid.*, p. v). His viewpoint is thoroughly psychologistic, but the psychology he preaches is *physiological*. The sensory qualities of sight and sound, of taste, smell, and warmth, he argues, are just movements in the nervous system. Physiologists like Lotze and Helmholtz assume them to be qualitatively different from such movements, 'but they usually consider this as self-evident without analysing and comparing the relevant concepts, in spite of the immense significance of the question' (*ibid.*, pp. 18–19). Consciousness is just reflexive movement in the fibers of the brain. 'It is made possible by the construction of the brain' (*ibid.*, pp. 27–8). What we call the association of ideas is just a case of resonance in the brain (*ibid.*, pp. 43–4). The soul or self is just the sum of such psychological–physiological activities (*ibid.*, p. 98).

These views are bound to have radical implications for the way one understands human thinking, concepts, and the principles of logic and mathematics. And these implications are of particular interest in the present context. All concepts arise out of sensory perceptions for Czolbe; there are no *a priori* concepts (*ibid.*, p. 53). He finds himself in agreement on this matter with Comte and Mill. But concepts do not, according to him, come about in quite the manner described by the empiricists— through comparison and abstraction. Rather, they originate 'in a purely physical process which under all circumstances is independent of the will' (*ibid.*, p. 64). Concepts, judgments, and inferences come out of perception 'with physical necessity' (*ibid.*). Our ability to form concepts is due to the molecular structure of the brain (*ibid.*, p. 52). An inference results physically from its two premises (*ibid.*, p. 58). Czolbe also rejects the Kantian doctrine, which he finds in the otherwise congenial logic of Drobisch, that concepts originate in judgments (*ibid.*, p. 55). A judgment is rather a composite of previously constructed concepts.

The principle of contradiction also arises out of sensory perception. It is a mistake to regard it as innate. The same is true of the principles of identity and excluded middle (*ibid.*, p. 60).

> The doctrine of innate concepts and laws of thought — which are in their original suprasensual (speculative) — . . . lacks any sufficient reason and has lately also been severely shaken by Comte, J. Herschel, Mill, Drobisch, Opzoomer and others (*ibid.*, p. 61).

The axioms of mathematics equally have an intuitive basis as Mill and others have seen. Although mathematics is an abstract science, its objects are definitely sensory or intuitive (*ibid.*, p. 39), not suprasensual and speculative as many mathematicians hold without *any* justification (*ibid.*, p. 38). Space and time, which were considered *a priori* intuitions by Kant and as such the foundations of the *a priori* truths of mathematics, are nothing but abstractions from sensory perceptions (or ideas) and their properties (*ibid.*, pp. 107ff).

Although Lotze had unwittingly helped to foster Czolbe's materialism, he felt it necessary to respond to this 'positive refutation' of his views. He criticized it for its unconvincing materialistic analysis of sensations, for its failure to realize that thinking always adds something suprasensual to intuition and is thus to be distinguished from it, and for its realist assumption that things are given to intuition, whereas in fact the concept of a thing is *a priori* and suprasensual.

Had Czolbe really shown that sensations, experience, and thought are identical with movements of the brain and nervous system? He had called consciousness a reflexive movement, but, Lotze wrote, 'admitted that all consciousness possesses the character of reflexivity, its total essential nature does surely not rest in this formal predicate alone.'[33] And later, after Czolbe had tried to clarify this point, he added: 'Only in so far as this circular movement is a circle of knowledge could it lead to self-consciousness; in so far as it is only a circular motion of atoms it would remain a completely external happening whose importance for the origin of self-consciousness need not be completely denied, but must be restricted to the claim that out of it alone consciousness could not arise.'[34] Lotze thus raises one of the major considerations at the heart of the continuing struggle between the defenders and the detractors of the identity-thesis. He was surely right in believing that Czolbe's particular defence of the identity-thesis was open to question on both physiological and philosophical grounds.

Czolbe himself, under the pressure of such objections, came to adopt the view that material objects and sensations form two mutually irreducible categories of entity, but that the assumption of such a duality was compatible with the principle of sensualism.[35] The issue was indeed embarrassing for the scientific materialists generally, and not only for Czolbe. For it brought out the tension that existed between their materialistic realism and their professed empiricism. The question was: if all thinking and knowing of the physical world arises out of sensations and if sensations cannot be explained in materialistic terms, what reasons are there for assuming a material world at all? Was a consistent naturalism forced back to the assumption that only sensations were given and that the material world is a construction out of them? In its later phase (i.e., after 1870) scientific materialism began to give way to phenomenalism, and realism to subjective idealism, all this on the basis of the rejection of the suprasensual.

Frege commented on this fact in 1879 (NS, pp. 155–6):

> Psychological treatments of logic have their source in the error that the thought (the judgment, as one usually says) is something psychological just like an idea. That leads then necessarily to epistemological idealism. . . . This flowing into idealism is the most remarkable in physiological psychology, because it is in such sharp contrast to its realistic starting-point.

He goes on to describe how this position, in its desire to be scientific, argues from nerve fibers and ganglion cells as objectively given. 'But it does not stay there: one also proceeds to thinking and judging, and then the initial realism suddenly changes into extreme idealism. . . . Now everything dissolves into ideas and all the earlier explanations thereby become illusory' (*ibid.*, p. 156). Frege's implication, like Lotze's, is clearly that both materialism and phenomenalism, both realism and subjective idealism, are unstable, unsatisfactory positions. What is needed to escape from them is the recognition of the suprasensual element in thinking.[36]

In his critique of Czolbe, Lotze turns for his second counter-argument to an examination of the difference between intuition and thought. His thesis is that 'thinking . . . consists in the addition of the suprasensual to intuition.'[37] The thesis is clearly akin to the Kantian conception of the relation of thought and intuition. The suprasensual, Lotze says, produces the only means of bringing the multiplicity of our intuitions into theoretical connection (*ibid.*, pp. 241–2). All theorizing presupposes it. The science of mechanics is not derivable from sensory, intuitive qualities alone, but only *from the thought* that these qualities are the predicates of a real subject (*ibid.*).

> Thinking does not consist in a chasing of images, not in a succession of ideas which are only intuitively connected or separated; no concept is expressed without the presupposition of an inner connection of its marks. . . . No judgment consists in a mere combination of subject and predicate; the copula possesses everywhere the sense of an inner, thoroughly non-sensory nexus which justifies their connection (*ibid.*, p. 246).

As late as 1918 Frege was to repeat Lotze's sentiments (KS, p. 360):

> Having impressions is necessary for the seeing of things, but not sufficient. What has to be added is nothing sensory. And nevertheless that is precisely what opens up the world for us; without this non-sensory element everyone would remain locked into his own inner world.

For Lotze such considerations undermine realism, whether of Czolbe's variety or any other. There is no intuitive, sensory, empirical concept of a thing. Realism assumes that things present themselves to our senses, but that is surely mistaken.[38]

> We see certain sensory qualities gathered in points of space, but their inner link which motivates not only their being together, but also their belonging together, can certainly not be observed intuitively. Whenever we call something a thing, we add something suprasensual to the content of our intuition.

This theme is also taken up by Frege. The above passage from 'The Thought' in fact appears in a context in which Frege argues that objects are not presented to us by the senses but are the product of the combination of sensory qualities with non-sensory elements, the non-sensory element here being the concept of an object. It is an anti-realist theme familiar enough from Kant, for whom the concept of a thing is also a formal, rather than an empirical, concept.

8 *The decline of naturalism*

Throughout the period of its dominance, roughly from 1830 to 1870, naturalism had not remained undisputed. It had been hotly attacked by Christian believers, by those who saw it allied to political radicalism, and by philosophers who had tried to preserve the traditional philosophical heritage. Lotze was a representative and outstanding figure in this last group. While recognizing the philosophical significance of the results and precise methods of the natural sciences, he insisted on the possibility of a speculative, idealistic metaphysics. But such a metaphysics, he argued, could not be constructed *deductively*, from first principles, as the German idealists of the beginning of the century had tried to do. It had to respect the autonomy of the sciences and to proceed *inductively* to its speculative conclusions. Philosophy, in other words, had to recognize and emulate the spirit of the sciences without itself dissolving into science.[39]

Lotze's writings reveal clearly why the counter-attacks of traditional philosophy against the naturalistic tide persisted and were nevertheless so ineffective. They are full of learning, subtle distinctions, and arguments, but they lack the vigor of new insights or discoveries. Lotze's philosophy aims at a syncretism of realism and idealism and tries to model itself in a backward-looking way on the thought of Leibniz and Kant. Lotze is forever reconciling unreconcilable opposites—scientific method and aesthetic feeling, modern science and Christian faith. The style of his writings as well as their content, for all their obvious philosophical sophistication, have for us the musty smell of a ponderous Victorian piety.

Lotze's writings do show that the existence of a dominant philosophical tradition is not incompatible with the contemporaneous existence of other philosophical traditions in more reduced circumstances.[40] The picture of closed historical periods characterized by unified systems of ideas which has often guided historiography is merely a product of deficient historical vision[41] and a speculative, Hegelian cast of mind that forever seeks to unify a multiplicity of diverse phenomena into a single pattern.[42] There is really no reason on earth why datable historical periods should be single intellectual units. Through the period of naturalism's dominance other and older philosophical traditions also survived. Eventually, political and social, scientific and philosophical reasons combined to bring the naturalistic period to an end, just as their combination had ended the rule of German idealism. At that point the surviving older traditions could provide the basis for a new start.

The transformations of German philosophy in the nineteenth century are without exception tied to political and social transformations: the decline of Kantianism is linked to the effects in Germany of the French Revolution and the Wars of Liberation against Napoleon; the dominance of idealism to the period of restoration before 1830; the naturalistic rejection of speculative thought to the revolutions of 1830 and 1848. By 1870 the radical political spirit of the mid-century had begun to subside in Germany. The country was ready for unification under conservative, Prussian, small-German auspices. These political developments reflected a shift in social attitudes, not least among intellectuals. Whereas the naturalists had been politically and socially active and often inclined to radicalism, the new generation was more moderate and less activist in its views. In academic circles conservative, nationalistic, even reactionary views became acceptable, sometimes with newly reawakened anti-Semitic overtones. From the little we know of Frege's political views it seems that they were pretty much in accord with those of his generation. The social and political reaction against naturalism was of course not confined to Germany, but spread throughout the Western world. One need recall only the long and hard struggle on which the Roman Catholic Church embarked at that time against what it called 'modernism'—that is, the forces of materialism, socialism, and democracy. In the English-speaking world this development had its counterpart in a renewal of religious fervor and in the growth of High Church piety; in philosophy the new age awakened an interest in idealistic philosophy and in Hegel in particular.

Trust in the power of naturalistic modes of explanation had begun to wane by 1870. Could the materialists account for mental activity or were their claims merely rhetorical?[43] Was science really derived from experience, or did the use of highly abstract concepts in mathematical physics show the connection between theory and experience to be much more tenuous than the naturalists had supposed? If knowledge was the function of a purely natural organism which had developed in the way the Darwinians described it, then were not knowledge and science determined by the structure and needs of the organism rather than simply by the external objects of knowledge, as was thought on the realist view? Were not the physiological, psychological, mathematical, and logical doctrines of the materialists of the middle of the century hopelessly out of date? And could the materialists, with their physiological background, really talk coherently about the historical and social sciences at all?[44]

Faced with such questions, naturalism did not just suddenly fade away, but it began to lose its intellectual respectability. For a while it remained in the popular imagination. In Germany its great propagator became Ernst Haeckel, who taught at the University of Jena at the same time Frege was active there. Haeckel's lectures, unlike Frege's, enjoyed a tremendous public interest. His books were widely read throughout Europe. Most successful among them was the *Welträtsel* [The riddles of the universe] of 1899, which was easily the most popular philosophical treatise of its age. In it Haeckel proposed solutions for all the major philosophical problems that had beset mankind; in each case, it turned out, a Darwinistically inspired monism was the solution. But, for all his popularity, Haeckel was no longer considered a serious threat by philosophers.[45]

II

Philosophical Reconstruction

1 New movements of late-nineteenth-century philosophy

In 1874 Franz Brentano introduced himself as the new Professor of Philosophy at the University of Vienna with a lecture on 'The Reasons for Disenchantment in the Field of Philosophy.'[1] He began by complaining that philosophy no longer enjoyed widespread intellectual trust. 'Almost universally it is thought to have as its chosen goal a veiled picture through whose cover no human eye can penetrate. . . . Most believe that philosophy cannot properly be counted among the sciences. They prefer to put it next to astrology or alchemy' (ibid., p. 86). Where philosophy lecture-rooms had formerly been crowded, they were now depressingly empty. Where the subject had once been a central part of the culture, it had now become completely marginal. What were the reasons for the decline? Brentano listed four: 'A lack of generally accepted theorems; the total revolutions which philosophy suffers again and again; the empirical unattainability of its chosen goal; and the impossibility of practical applications' (ibid., p. 92). These had of course been characteristic features of philosophy since its beginning. They had repeatedly led to scepticism about the claims of philosophy and even about the possibility of any reliable human knowledge. Since the seventeenth century the advances of the empirical sciences had thrown the perturbations of philosophy into an unfavorable light. But, despite the general distrust, 'there remains for philosophy a circle of questions whose claim to be answered must not and, for the sake of humanity, cannot be given up. She has therefore indubitably a place to fill among the sciences and a future is assured for her' (ibid., p. 98). According to Brentano, this circle of philosophical questions is to be found in psychology, which cannot be exhaustively explored with the methods of the empirical sciences. 'Signs are not lacking to indicate a period of philosophical awakening to a fruitful life. The conditions are given; the method is prepared; the research procedure has been demonstrated' (ibid., p. 99).

PHILOSOPHICAL RECONSTRUCTION

Brentano's lecture reflects very clearly the shift in the philosophical climate of the 1870s. The intellectual dominance of naturalism was coming to an end and post-naturalistic philosophers had begun to explore new ways of philosophizing. They wanted philosophy to be scientific; that is, they were willing to acknowledge the achievements of the sciences and the success of their empirical methodology and they were willing to grant that philosophy could learn from the conscientiousness and precision of scientific procedures. Beyond that they did not always agree on what it meant for philosophy to be scientific, but they all believed that philosophy had a function different from that of the empirical sciences and that it would therefore not dwindle away into them.

The naturalists had shared with the German idealists a belief in the unity of human knowledge and both had concluded from this that there could be no clear-cut boundary between philosophy and science. The post-naturalists, on the other hand, agreed with the Kantian assumption of a sharp dividing-line between the two. They considered themselves critical thinkers, concerned with establishing the limits of empirical knowledge. They tended to think of the naturalists as epistemologically naive because of their realism and materialism. The questions they raised about the limits of human understanding, about the contribution that human forms of understanding make to scientific theorizing, and about scientific theories as pictures of reality, all drew them back to Kant's critical philosophy. The cry 'Back to Kant' had first been voiced in 1865, but as the century went on it became louder and more powerful.[2]

Of particular significance for the post-naturalist attempts to separate philosophy from empirical science was a distinction first made by Kant, but which now gained a new life. It is the distinction between the subjective psychological conditions of thinking and the objective contents of such empirical acts. Brentano's claim that psychology could not be exhausted by empirical means was based on drawing that distinction. The contents of thought, he argued, required a new, descriptive kind of psychology. Husserl's phenomenology and Frege's pure logic equally depended on it.

Three groups of philosophers demand our attention in connection with Frege's thought: (1) the Neo-Kantians, (2) the critical positivists, and (3) the phenomenologists.[3]

Neo-Kantians were ready to grant that a successful scientific theory would probably have to be cast in mechanistic and materialistic terms, but they rejected the naturalist claim that this showed the truth of materialist ontology. Human cognition could not be reduced to material processes. The naturalists had overlooked the fact that their materialism depended *on a theory*, that is, something that was itself a product of human cognition. The materialist doctrine could therefore not provide a full account of human understanding. Like any other theory it made

36

certain *a priori* assumptions about the nature and powers of the human mind.[4]

The impact of this epistemological critique of naturalistic materialism was so great that Neo-Kantianism remained the dominant school of philosophy in Germany for more than a generation.[5] One of the strongholds of the school was Frege's own university, Jena. It could look back on a very distinguished connection with the German idealists and it was duly proud of that link.[6] From 1882 to 1911 Otto Liebmann, the originator of the slogan 'Back to Kant,' had taught at the university. Later on, Bruno Bauch, who had connections with both the Marburg and the Heidelberg school, came to Jena, where he maintained close contact with Frege and was also the supervisor of Carnap's dissertation.

For all their visibility and the large volume of their writings, the Neo-Kantians did not really succeed in creating new forms of philosophizing. They made probably their most important contribution in the interpretation of the great philosophers of the past, in particular, of course, Kant. By 1930 the movement had reached the end of its creative life and quickly faded away. The students of the last Neo-Kantians became phenomenologists, Heideggerians, or logical empiricists.[7]

The critical positivists, foremost among them Mach and Avenarius, were in certain respects closer to the naturalist tradition than the Neo-Kantians or any other of the post-naturalist philosophers. Nevertheless, their positivism must be clearly distinguished from the dogmatic positivism of the middle of the century (just as it must be distinguished from logical positivism). Unlike the dogmatic positivists (that is, Comte and his followers) and unlike the scientific naturalists, the critical positivists were neither materialists nor realists. Like the other post-naturalists they were enough influenced by Kant to consider such a view naive.[8] According to Mach scientific theories are never explanatory; they are merely economic devices for summarizing and organizing experience. The naturalist interpretation of science was in his eyes only a residue of illegitimate metaphysics.

Mach and the other critical positivists agreed with the naturalists, however, in considering human thinking to be a merely empirical, natural phenomenon. It was precisely this conception that formed the basis of their pragmatic, economic view of scientific theorizing. For Mach 'the demands of economy go further than those of logic.'[9] Language is also 'an economical contrivance' (*ibid.*, p. 578). Mathematics, pursued without Mach's economical theory in mind, 'is scarcely of more educational value than busying oneself with the Cabala. On the contrary, it induces a tendency toward mysticism, which is pretty sure to bear fruits' (*ibid.*, pp. 585–6). The formation of the concept of number, when psychologically examined, 'is just as much initiated by experience as the formation of geometrical concepts.'[10]

In Mach's conception logic would be no more than a 'universal economy of thought.'[11] And this idea was clearly not too far from the naturalistic interpretation. Understandably, Husserl therefore rejected Mach's views as psychologistic, claiming that they could throw no light 'on the ideal laws of pure logic.'[12] Wilhelm Jerusalem, one of Mach's associates in Vienna, responded with a sharp attack on Husserl's idea of a pure logic which could just as well have been directed against Frege's similar ideas. Interpreting logical laws in Mach's sense as 'thought-economical rules,'[13] Jerusalem writes (*ibid.*, p. 102):

> We consider ourselves also a piece of nature and believe, therefore, that the laws according to which our intellectual life develops and regulates itself are also natural laws. For that reason we assume that the mathematical and logical laws are also natural laws and that they are knowable not a priori, but only through experience.

Jerusalem attacks not only Husserl's belief that the logical laws are *a priori*, but also the assumption that they are truths in themselves, i.e., that they would remain true even if no man recognized or could recognize their truth. And, finally, he also rejects Husserl's assumption that in real, subjective acts of judging we can separate out an ideal, objective content which has that property of being true in itself and *a priori*. According to Jerusalem, Husserl's position involves a misconceived attempt to revive scholastic ideas whose ultimate source lies in Pythagorean and Eleatic thought. In contrast, he considers the psychological view of the logical laws 'the greatest progress which logic has made since Aristotle' (*ibid.*, p. 78).

The position of Mach and his followers shows that psychologism was becoming a broader movement in the last quarter of the nineteenth century. It was no longer tied exclusively to the realism and materialism of the scientific naturalists. For Mach the ultimate givens were not material objects, but sensations which could be interpreted alternatively in mentalistic or physicalistic terms. Nevertheless, the ultimate explanation of the logical laws was to be sought in 'the difficult study of the evolutionary history of man' (*ibid.*, p. 95). Other late-nineteenth-century psychologists, such as Erdmann, went further and adopted a full-blown phenomenalism. As the discussion proceeded, psychologism became identified simply with the thesis that the logical laws are empirical, natural laws, without any assumptions being made concerning the nature of the empirical. Thus, Husserl could eventually attack it as just a version of radical empiricism.[14]

Mach's psychologism was rejected not only by Husserl but also by the logical empiricists of the next generation, despite the fact that they considered themselves the direct descendants of Mach's thought. They accused Mach of an 'underestimation of the role of mathematics and logic in the building of science,'[15] and because of this disagreement did not pursue any further Mach's thought-economical interpretation of science. In spite of their appeal to the name of Mach, it is therefore open to doubt whether the logical empiricists can really be considered his successors. Their disagreement with him is of course due to the fact that the logical empiricists, influenced by Frege, Russell, and the early Wittgenstein, had come to reject radical empiricism and insisted on a sharp distinction between analytic and synthetic, and between logical and empirical, truths. Their views were therefore a good deal more rationalistic and traditional than those of Mach and his immediate followers.

Among those followers our attention is drawn to Fritz Mauthner, who had studied with Mach in Prague and remained in lifelong contact with him.[16] In Mauthner's hands, under the influence of Nietzsche's reflections on language[17] and the literary criticism of Otto Ludwig, Mach's critical epistemology became transformed into a critical philosophy of language.[18] Just as Mach and Jerusalem were trying to understand human theorizing in terms of a naturalistic account of experience, Mauthner tried to understand it in terms of a naturalistic account of language. Mach and Jerusalem themselves recognized the relevance of Mauthner's considerations to their own undertaking.[19] In accordance with the naturalistic assumptions, Mauthner's concern was entirely natural language. Mach had argued that mathematical physics could not reveal to us the real nature of things; Mauthner similarly reasoned that a formal calculus could not reveal the real nature of language, that it could only give us at best a practical device for dealing with language. Mauthner's philosophy of language is thus always and only a philosophy of ordinary language. His writings show that what is known as psychologism is not necessarily in conflict with a belief in the fundamental role of language in human understanding. That would be so if psychologism meant the same as subjective idealism; but the term has in fact a much wider meaning and refers to all those philosophical movements which try to account for human understanding exclusively in empirical, naturalistic terms.[20]

Husserl, too, began his philosophical thinking from psychologistic premises. His attempt to provide foundations for arithmetic in his *Philosophie der Arithmetik* of 1891 was conceived in an entirely psychologistic spirit. At the time he had thought he could show the untenability of the anti-psychologist views Frege had expounded a few years earlier in his *Foundations of Arithmetic*. Husserl's work provoked a sharp response from Frege in a review published in 1894 (KS, p. 192):

> When reading this work I came to recognize the devastations which have been brought about by the incursion of psychology into logic and I have considered it my task to bring the damage fully to light. The

errors which, in my opinion, needed to be exposed are due less to the author than to a widespread philosophical disease.

Two years later Husserl had completely abandoned his psychologism.[21] In spite of the fact that his views were now closer to Frege's than to any other contemporary writer, he could not bring himself to acknowledge Frege's influence on the change in his views in more than a grudging footnote (*ibid.*, p. 169). He made no mention of Frege's review, though it easily constituted the most sustained attack on psychologism before Husserl's own book. Even many years later Husserl could refer to Frege only as a man who was generally considered 'a sharply intelligent outsider [*Sonderling*] who was bearing fruit neither as a mathematician nor as a philosopher' (WB, p. 92).

While he failed to mention Frege, Husserl did refer explicitly to the influence of Bolzano and Lotze (*Logische Untersuchungen*, p. 227). In rejecting his own earlier psychologism, Husserl was aware that he was setting himself in opposition to a prevailing opinion (*ibid.*, p. 3, also pp. 211–12). He considered his move a deliberate 'link back to justified tendencies of older philosophy' (*ibid.*, p. 213). In Kant's separation of pure and applied logic, in Herbart's insistence that conceptual contents are objective and must be distinguished from subjectively entertained thoughts, in Lotze's *Logik*, even in the writings of Neo-Kantians like F. A. Lange he saw tendencies similar to his own. He felt particularly close to Leibniz (*ibid.*, p. 219) because of his interest in a universal mathematics as an extension and foundation of logic and because his interest in the problem of logical notation made him the precursor of mathematical logic. It was Leibniz's rationalism and his rationalistic conception of language that Husserl found appealing. Finally, he claimed to recognize in Bolzano 'one of the greatest logicians of all time,' stressing his mathematical precision and philosophical sobriety.

> Historically, he is to be put into fairly close relation to Leibniz with whom he shares important and basic ideas and to whom he is also philosophically closest. Even he did not of course exhaust the wealth of Leibniz's logical intuitions, in particular not with respect to a mathematical syllogistic and the *mathesis universalis* (*ibid.*, p. 225).

I list the philosophers Husserl cites in connection with his own antipsychologism to show how close those sources are to those that influenced Frege. For the great philosophical influences on Frege were no doubt Kant and Leibniz, Lotze and Herbart. With respect to the first three that influence can be readily uncovered, and occasional references to Herbart seem to reveal Frege's indebtedness to him. Frege does not explicitly refer to the Neo-Kantians or to Bolzano, but it is obvious that his ideas are akin to theirs.

Because of the kinship between Frege's views and those espoused by Husserl after his break with psychologism it has been said that Frege's 'domain of the objective unreal can be nothing else but the domain of phenomenology.'[22] The comparison is inviting, but because of the difficulty of Husserl's views and the insufficiently understood historical relationship between them, it falls outside the scope of the present study.

In any case it is clear that important differences remain between their respective positions. Frege's fundamental concern is with the construction of a logical language in which objective conceptual contents can be represented. The examination of objective conceptual contents is the task of logic, and logic is fundamental for Frege. While Husserl can grant that metaphysical and epistemological considerations must be founded on logic, he ultimately considers phenomenology 'understood as a pure theory of the essential elements of experience' as even more fundamental.[23] Phenomenology, understood in Husserl's sense, is moreover a descriptive rather than a discursive discipline. While Husserl draws attention to the formal languages of mathematics and mathematical logic, his own concern is in the end a phenomenological description carried out in everyday language.

2 Frege's foundational interests

Frege was born in 1848, the year of the great democratic revolution; but the echoes of the turmoil shaking Germany must have been faint in his birthplace. Wismar was a small market town in Pomerania, far from the big cities and their political and social ferment. The family he was born into belonged to the staid Lutheran middle class of the region, ethnically mixed (partly German and partly Polish) and with some remaining ties to the land. The father, Alexander Frege, owned and directed a school for young ladies, which his wife took over after his early death.

When Gottlob was twenty-one his mother sent him to the small but renowned University of Jena. At that time he seemed to have already decided on mathematics as his major field of interest, but he also took courses in chemistry and philosophy (the latter from Kuno Fischer, who was soon to leave for Heidelberg). It is possible that he was preparing himself to take over the directorship of the school from his mother. One of his teachers at Jena, the mathematician and social reformer Ernst Abbe, quickly recognized the talents of the young student and became his faithful supporter throughout his life. It is almost certain that Frege would not have succeeded in obtaining an academic position at Jena without Abbe's help. On Abbe's advice he left Jena to obtain his Ph.D. under the mathematician Schering at Göttingen. There he attended the lectures of Lotze, whose thought was to have a permanent influence on him.

PHILOSOPHICAL RECONSTRUCTION

After he had received his doctorate at Göttingen, Frege returned to Jena and with Abbe's assistance qualified for the position of *Privatdozent* in 1874. Five years later, after publication of the *Begriffsschrift*, the first exposition of his new logic, he was promoted to *ausserplanmässiger Professor* again on Abbe's recommendation.

In order fully to understand what led him to conceive his new logic we must take one step backwards and look at his interests and concerns in the period before 1879. At the end of his life Frege himself characterized his initial motivations in these words: 'With mathematics I began. There seemed to me a most urgent need for better foundations in that science. . . . The logical imperfection of language was an obstacle for such investigations. I sought a remedy in my *Begriffsschrift*. Thus, I came from mathematics to logic' (NS, p. 273).

Since its beginnings as a science in the Pythagorean circle mathematics has given rise to philosophical and foundational questions. It has been the source of ontological, epistemological, and methodological reflections. The Greeks asked: What and where are numbers? Why is the natural world countable and measurable? How do we know the truth of mathematical propositions? What is the proper method of proof and definition in mathematics? It is not just that mathematics stimulates philosophical questioning, but philosophy also seems to determine what kinds of mathematics are permissible. The Pythagorean has mathematical problems with the infinite which are different from those of the Aristotelian. The objectivist (or Platonist) allows himself a stronger kind of mathematics than does the subjectivist (or constructivist). The Kantian permits Euclidean but rejects non-Euclidean geometry.

Given Frege's belief in 'the urgent need for better foundations' in mathematics, we should ask what kind of need he perceived and where he hoped to find better foundations. It turns out that his considerations are closely connected to the foundational issues that other mathematicians and philosophers were facing in nineteenth-century mathematics.

Mathematics played a double role in the development of post-naturalist thought. During the course of the century mathematics was applied more and more extensively to scientific and technological problems.[24] This mediating role of mathematics undermined the naturalistic (one can also say empiricist) conviction that science was based directly on experience and was derived from it by simple inductions, i.e., abstraction and generalization. The other role of mathematics in the formation of post-naturalist thought was due to the internal theoretical development of mathematics itself. Not only the empirical sciences were growing rapidly during the century; the same was true of mathematics. And this growth generated a wealth of philosophical problems.

It has been said that mathematics became almost a new science after Gauss. From a relatively informal discipline at the beginning of the century, it turned into a substantially more rigorous, abstract, and systematic science. The continuing problems with Euclid's fifth postulate (the parallel postulate) led Gauss to conceive the possibility of non-Euclidean geometries, which were developed further by Riemann, Bolyai, and Lobatschewskij. This led in turn easily to the strictly axiomatic treatment of geometry by Pasch and the formal geometry of Hilbert. The analysis of the infinite had induced Bolzano to investigate the paradoxes of infinity; in Frege's time it generated the theories of irrational numbers of Dedekind and Cantor and related reflections on the concept of a function. And out of these developments in turn came Cantor's theory of point-manifolds and, by generalization, his theory of sets. The theory of natural numbers, on the other hand, was at first considered unproblematic; its foundational problems were left to Peano, Frege, and Russell.[25]

These developments brought forth a whole generation of philosophizing mathematicians. Some of them tried to preserve a strictly naturalistic interpretation of the science (among them Helmholtz, Kronecker, and the early Husserl), while others (Frege himself, Dedekind, and Cantor) considered the new developments in mathematics compatible only with an anti-naturalistic philosophy. In his review of Cantor's collected essays on the theory of the transfinite, which was for the most part quite negative,[26] Frege wrote:

> I do not want to conclude this review without repeating one sentence with complete agreement: 'Thus we see that the academic-positivistic scepticism which is now dominant and mighty in Germany has finally also reached arithmetic, where it seems to be drawing its last possible conclusions with the utmost consistency and will perhaps destroy itself.' Indeed, here is the cliff where it will founder. For the infinite after all cannot be denied in arithmetic and yet it is incompatible with that epistemological attitude. Here, it seems, is the battlefield where a great decision will be reached (KS, p. 166).

The urgent need for better foundations in mathematics was for Frege first of all a need for philosophical reassessment. His purpose was to show that mathematical formulas are not empty symbols, that they are not inductive generalizations, nor expressions of psychological laws of human thinking. He was opposed to 'those empirics who recognize induction as the sole original process of inference (and even that as a process not actually of inference but of habituation)' (F, p. xi). Such views were to be defeated by the definitive proof that mathematics consists of truths and that those truths are *a priori*. What motivated Frege was his hope of completing and correcting the Kantian enterprise: to reveal once and for all the truth of Kantian apriorism, to show that Kant's conception of the nature and limits of human understanding is essentially correct and that truth is objective, not just a subjective-psychological illusion.

Such a clarification of the philosophical foundations of mathematics, he hoped, would at the same time show where and how future work in mathematics should be done. With the foundations secured, mathematics could truly become the science it was meant to be.

3 Frege's pre-Begriffsschrift writings

The initial motivation of Frege's concern with the foundations of mathematics (and implicitly his concern with logic) seems to be epistemological in character. Michael Dummett has given another interpretation of the historical situation. In his *Frege* he writes: 'Descartes' revolution was to make epistemology the most basic sector of the whole of philosophy. . . . Descartes' perspective continued to be that which dominated philosophy until this century, when it was overthrown by Wittgenstein, who in the *Tractatus* reinstated philosophical logic as the foundation of philosophy' (p. xv). And he adds that, though Frege did not explicitly discuss this issue, 'by his practice he demonstrated his opinion that logic could be approached independently of any prior philosophical substructure' (*ibid.*).

It is of course true that Descartes had spoken contemptuously of logic and had considered his own epistemological considerations a separate issue, and it is also true that for the empiricists epistemological problems were problems of perception and of the nature of mental processes rather than purely logical problems, but it is quite wrong to find a general conflict between logical and epistemological considerations. Leibniz, as has been said, had criticized the Cartesian separation of the two. And later on Kant saw epistemology and logic as intertwined. Frege's interest in logic shows not a lack of concern with epistemology but rather a particular anti-empiricist (or anti-naturalistic) epistemological viewpoint.

The nature of this epistemological concern is clearly revealed in Frege's writings before his *Begriffsschrift*. There are two of them, his doctoral dissertation and his *Habilitationsschrift*. The topic of the doctoral dissertation ('On a Geometrical Representation of Imaginary Figures in the Plane,' KS, pp. 1–49) reflects Frege's persistent interest in problems of geometry. This is an interest he seems to have had even before he was especially interested in arithmetic and logic. The extent to which Frege was concerned with geometry throughout his life does not become fully apparent from his major writings. It was not only an interest in foundational issues (cf. KS, pp. 90–1, 94–8); Frege regularly lectured on geometry and his notes show him concerned with various technical geometrical problems.[27] It seems that at one time he was considering a 'new geometry' to complement the new arithmetic of the *Grundgesetze*.[28]

The doctoral dissertation begins with the fundamental claim that geometry rests on intuition. Now this claim could be taken in two quite different senses. On the one hand, it could be considered a restatement of the Kantian thesis that 'geometry is a science which determines the properties of space synthetically, and yet *a priori*' (*Critique of Pure Reason*, B 40). Kant elaborates on this thesis by asking: 'What then must be our representation of space, in order that such knowledge of it may be possible? It must in its origin be intuition. . . . Further, this intuition must be *a priori*, that is it must be found in us prior to any perception of an object, and must therefore be pure, not empirical intuition' (*ibid.*, B 40–1). Frege's claim about the intuitive character of geometry might also be regarded as an appeal to *empirical* intuition as the basis of geometry. Earlier in the century Gauss, in realizing the logical possibility of non-Euclidean geometry, had argued that the validity of the geometry of actual physical space would have to rest on just such an empirical intuition.[29] Since Frege had worked in Göttingen with Ernst Schering, the editor of Gauss's works, we may assume that he was familiar with this possibility of claiming an intuitive basis for geometry. But it is obvious from his actual formulation that it is the Kantian viewpoint he is expressing and not Gauss's. What he *says* is that geometry rests 'on axioms that derive their validity from the nature of our capacity for intuition [*Anschauungsvermögen*]' (KS, p. 1) — that is, from the *nature* of a *capacity*, rather than from particular intuitions.

The issue is not merely of epistemological interest, for it carries with it ontological implications. 'The question touches immediately on metaphysics,' as Gauss put it.[30] For if it is empirical intuition that decides which geometry is valid, then, according to Gauss, 'we must humbly admit that even though number is only a product of our mind, space also possesses a reality outside the mind to which we cannot ascribe its laws *a priori*.'[31] In other words, the issue is between Kant's conception of space as transcendentally ideal and a realistic conception.

Michael Dummett has interpreted Frege as a realist in revolt against a dominant idealism. Unfortunately the concept of realism has many different uses. There certainly seems to be at least one sense in which Frege was not a realist. Unlike Gauss, and unlike the mid-century naturalists, Frege held a Kantian view of space and hence a transcendentally subjective view of the objects that occupy it.

Frege never abandoned the conception of geometry as synthetic *a priori*. It remained one of the stable elements in his thought. It was reaffirmed in the *Foundations of Arithmetic* (e.g., pp. 100–1) and again in his very last notes (NS, pp. 292ff, 298ff). When he had convinced himself that secure *a priori* foundations for arithmetic could not be discovered in logic, he naturally turned to geometry to provide them. In accordance with this

Kantian conception he kept saying that 'non-Euclidean geometry must be reckoned among the non-sciences, which, as mere historical curiosities, one deems worthy of scant attention' (NS, p. 184). He said this in full knowledge of Riemann's work, who had been one of the teachers of his own teacher Abbe. His opposition to Hilbert's formal geometry was not a late foible, but came from the very deepest level of his philosophical and mathematical thinking.

The detailed discussion in his doctoral dissertation springs directly from his Kantian viewpoint. The problem is how to understand the geometry of imaginary figures, since they often exhibit characteristics that seem to contradict intuition. In order to resolve the problem Frege suggests that the notion of an imaginary figure must be examined. The difficulties involved in the notion can be understood if we take the analogy of another difficult mathematical notion: that of an infinitely distant point. This concept also contains certain apparent contradictions. It is in fact a *contradictio in adiecto*. Nevertheless, the concept is perfectly admissible. One must understand that it is really an improper (*uneigentlich*) expression. The notion is used to signify the fact that parallel lines behave projectively like lines going through one point. An 'infinitely distant point' is what parallel lines have in common, namely, the same direction. The term constitutes only a convenient manner of speaking. Infinitely distant points can be represented by projecting a plane onto the surface of a sphere; this gives us an intuitive representation of such points. There is then no real difference between proper and infinitely distant points. The same kind of analysis, Frege suggests, can be given for the notion of an imaginary figure. That notion, too, is an 'improper' one and therefore in need of explication. We must look and see how the notion of an imaginary figure is *used* in geometry. We must interpret this non-intuitive notion in terms of intuitive ones. Only then can the proper content of the geometry of imaginary figures be brought out. Once he has described his program in this way Frege then sets out in his dissertation to provide the details of such an intuitive representation.

For Kant both geometry and arithmetic were synthetic *a priori* and based on pure intuitions. Gauss, on the other hand, had separated the two sciences, basing the one on empirical intuitions but declaring numbers 'a product of our mind' and arithmetical truths, presumably as a consequence, analytic. Frege's second dissertation, the *Habilitationsschrift* of 1874 ('Methods of Calculation which are Founded on an Extension of the Concept of Magnitude,' KS, pp. 50–84), parts company with Kant and joins the side of Gauss in its assessment of the nature of arithmetic. Frege writes:

There is a remarkable difference between geometry and arithmetic in the manner in which they justify their principles. The elements of all geometrical constructions are intuitions, and geometry points to intuition as the source of all its axioms. Since the axioms of arithmetic have no intuitiveness, its principles cannot be derived from intuition (KS, p. 50).

'Geometry,' as he says elsewhere, 'forms a transition from the material world to the world of the mental.'[32] Arithmetic, as the discussion of the second dissertation shows, belongs clearly to the world of the mental. For arithmetical propositions apply to magnitudes of any kind. Intuition can merely provide us with instances of their application, otherwise it is irrelevant. 'From this it follows that we put the propositions necessary for the construction of this science into the concept of magnitude' (KS, p. 51). The laws of arithmetic derive from a concept, not from intuition. 'We do not find the concept of magnitude in intuition, but create it ourselves' (*ibid.*).

Frege's problem is, therefore, how we can formulate this concept of magnitude to allow the largest possible number of applications. He first characterizes the notion of a field of magnitudes. Such a field of magnitudes is defined by a set of operations (or functions—Frege uses the terms indiscriminately) which take objects in the field as arguments. The field of natural numbers, for instance, is *one* such field of magnitudes and the operation of addition defines it.

> Quite generally speaking, the process of addition is the following: we replace a group of things by a single one of the same kind. This provides us with a clarification of the notion of sameness of magnitudes. If we can determine in every case when two objects coincide in a property, we obviously possess the correct concept of that property. Therefore, by determining under which conditions equality of magnitudes obtains, we determine the notion of magnitude (*ibid.*).

There are many distinct fields of magnitude. But the whole content of arithmetic is contained in the notion of magnitude, and specific kinds of magnitude, such as natural numbers or angles, can be defined from this standpoint. However, Frege says, the task of showing how this can be done would lead too far (*ibid.*).

The more specific task which he sets himself in the second dissertation is the definition of a concept of magnitude for operations (or functions). He had studied the theory of functions with Abbe and Schering and he now wished to show how his idea of a general theory of magnitudes of functions would relate various parts of arithmetic in a new way. The investigation is, however, only a beginning in his eyes. Because of the size and difficulty such a theory would have if it were developed 'in its greatest generality and completeness,' Frege promises only the elaboration of some immediate consequences of the idea (*ibid.*, p. 52). The rest is to be left to a later occasion. Abbe's reaction to the dissertation was that it might contain the seeds of a significant and comprehensive new treatment of analysis.[33]

What Frege had in fact argued in his second dissertation was that arithmetical propositions are analytic in the Kantian sense. Their truth follows from the concepts occurring in them; to be more specific, it follows from the concept of magnitude. One must distinguish two different claims here, namely:

(1) Arithmetical propositions are analytic;
(2) Arithmetical propositions are derivable from logical principles alone.

According to Frege's own later characterization (F, pp. 3–4), an analytic truth is one that follows from logical principles and definitions alone. Proposition (2) therefore implies (1); but the reverse is not necessarily the case. It holds only if we assume that the definitions of the arithmetical terms which are needed to derive all arithmetical propositions can ultimately be cast in purely logical terms. In the second dissertation, Frege assumed that arithmetical truths 'followed from the concept of magnitude,' but he did not raise the question whether this concept can be defined in a purely logical vocabulary. Proposition (2) expresses what is known as 'the logicist thesis' — the claim of the reducibility of arithmetic to logic. It seems then that in his second dissertation Frege assumed the analyticity of arithmetical propositions but had not yet raised in his mind the questions of the validity of the logicist thesis. In fact, no interest in questions of logic is visible in his first writings.

We must therefore ask what drove him to consider the logicist thesis. How did he move from mathematics to logic? He later gave this account of his own development: 'Soon I recognized that a number is not a heap, not a series of things; nor is it the property of a heap. A number statement which is made on the basis of counting contains an assertion about a concept. . . . Thus I came from mathematics to logic' (NS, p. 273). As he explained in the *Foundations of Arithmetic*, a number statement is an assertion about a concept because it is possible to look 'at one and the same external phenomenon' and say equally 'This is one group of trees' and 'These are five trees,' or 'Here are four companies' and 'Here are 500 men' (F, p. 59). But if number statements were assertions about concepts, their analysis would have to include an account of concepts. And that belongs to the area of logic.

4 Trendelenburg and the idea of a logical language

Frege's new interest drove him to the study of the logical literature. It must have been in the years between 1874 and 1879 that he read the works of the Boolean and philosophical logicians. Before 1874 he cannot

PHILOSOPHICAL RECONSTRUCTION

have had much use for their writings since there is no indication that he considered logic necessary for his foundational investigations, but, although the *Begriffsschrift* of 1879 contains only a few direct references, it shows him to be acquainted both with Boolean algebra[34] and with the writings of philosophers like Trendelenburg and Lotze.

Trendelenburg was one of the philosophers of the mid-century who had endeavored to keep traditional philosophy alive. Like others of his generation, he adopted a syncretist conservatism in philosophy which in his case — with his background as a scholar of Greek philosophy — was laced with a heavy dose of Aristotelianism. In his major work, the *Logische Untersuchungen*, he wrote: 'One must abandon the German prejudice that the philosophy of the future has still to find a newly formulated principle. The principle is already found; it lies in the organic world view that has its origin in Plato and Aristotle.'[35] In accordance with this precept he had set out to show how philosophy had to be founded on 'logic and metaphysics as the fundamental science' (*ibid.*, p. 4). The logic he had in mind was of course essentially Aristotelian, but the interpretation he gave it was a very special one. Aristotelian logic in Trendelenburg's sense is not identical with formal logic. 'The idea of formal logic originated with Kant and was received by Herbart in a Kantian sense' (*ibid.*, p. 35), but the Aristotelian form of logic is supposed to be quite different. 'Nowhere does Aristotle express the intention to understand the forms of thought only out of themselves' (*ibid.*, p. 30). He concerns himself with the psychological and metaphysical foundations of concept formation, of negation and assertion, of the concepts of necessity and possibility. Just as for Aristotle, logic is for Trendelenburg more than formal logic. At the center of his construction he puts the Aristotelian notion of purpose. In the *Logische Untersuchungen* he tried to reconstruct this logic of purpose and to incorporate into it the insights of Kant, Hegel, and Herbart. He also attempted to relate these logical investigations to the methodological and foundational problems of the individual sciences. For 'without careful attention to the methods of particular sciences, logic will fail its aim, for it has then no specific subject matter against which it can test its theories' (*ibid.*, p. iv).

It is possible that Frege read the *Logische Untersuchungen*, or at least parts of it. Trendelenburg's discussion of the relation of judgment and concept in the second volume, in which Gruppe's ideas on the subject are recounted and developed, may well have influenced Frege's thought.[36] It is, however, certain that Frege knew Trendelenburg's essay 'On Leibniz's Project of a Universal Characteristic.'[37] It seems likely that this essay first familiarized Frege with the Leibnizian idea of a logical language.[38] Frege certainly took one thing from Trendelenburg and that was the name for his notational system. Following Trendelenburg he called his logical symbolism a '*Begriffsschrift*,' a conceptual script.[39]

PHILOSOPHICAL RECONSTRUCTION

In his essay Trendelenburg describes Leibniz's logical projects, paying specific attention to his insight into the philosophical significance of language. In agreement with Leibniz, he writes:[40]

> Through the sign ideas which would otherwise be diffuse are separated and become, as separated elements, a permanent possession which the thinker can use. Through the sign distinctions are drawn, the distinguished is fixed, and the fixed becomes capable of new connections. Through the sign the idea detaches itself from the sensory impression . . . and now becomes capable of lifting itself up into the general. Thus, through the verbal sign, thinking becomes on the one hand free and, on the other, definite.

The classical German tradition, from Kant to Leibniz, had taken scant notice of language. Of the older philosophers only Leibniz had understood the point. His work in mathematics convinced him of the real importance of an adequate system of notation. In his early dialogue 'On the Connection Between Things and Words' he had already considered whether thought can exist without a language and had replied to himself:[41]

> Not without some sign or other. Ask yourself whether you can perform any arithmetical calculation without making use of any number signs. When God calculates and exercises his thought, the world is created.

Trendelenburg was fully familiar with Herder's and Gruppe's reflections on natural language, but in the essay in question his own interest was in the Leibnizian idea of the construction of a pure rational language. For both Leibniz and Trendelenburg natural language is a rather imperfect medium which for philosophical and logical purposes is in need of reconstruction. Language is indispensable for all 'inventions and discoveries, for all things which the human mind has acquired and constructed,' Trendelenburg writes,[42] but ordinary language realizes the task of the linguistic sign only partially. In it the connection between the signs and the corresponding idea is brought about *by association*. 'Only to a small extent is there an internal relation between the sign and the content of the signified idea' (*ibid.*). From this observation there arises the idea of a language which brings 'the shape of the sign in direct contact with the content of the concept' (*ibid.*). This kind of language is to be called a *Begriffsschrift*. It is partially realized in the symbolism of mathematics but even then Leibniz's goal is not yet fully achieved; there is not yet a language in which the connection between signs and concepts is *logical*, and not merely psychological.

Trendelenburg points out that the Leibnizian project has its origin in the speculations of Raymundus Lullus and the utopian projects of Kircher, Becher, Dalgarno, and Wilkins.[43] One might therefore consider his an equally fantastic enterprise. He recounts Descartes's critical remark on such undertakings to the effect that 'the invention of such a language depends on the true philosophy' (*Historische Beiträge*, p. 8). And he notes that Leibniz's project, as well as all the others, has remained incomplete since the necessary philosophical analysis of concepts has not yet been achieved. None the less, he considers Leibniz's idea not wild speculation, but a significant intellectual project. But its execution, as Descartes had seen, must be accompanied by a philosophical analysis.

The first step in the realization of the Leibnizian project must consist in scaling it down. Leibniz had tried to do too much too fast. Trendelenburg says that Kant's distinction between the content and the forms of human knowledge can be used to give the project a more realistic size. 'When Kant followed Leibniz and distinguished sharply between the form and the matter of thinking and made the complete knowledge of the mind-related form the essence of his critical philosophy, the realization of a characteristic that restricts itself to the formal side of thinking had come closer' (*ibid.*, p. 26). He points out that most Kantians had ignored the Leibnizian project, but at least one of Kant's followers, Ludwig Benedict Trede, set out in his *Proposals for a Necessary Theory of Language* (Hamburg, 1811) to construct a general characteristic 'on a Kantian basis in the spirit of Leibniz' (*ibid.*, p. 27). Trendelenburg considers Trede's work to be too closely tied to the Kantian categories, but his attempt to execute the Leibnizian project is nevertheless worthy of approval since Leibniz's original idea 'was the design of a wide-ranging and sharp mind' (*ibid.*, p. 29). Science will no doubt come to appreciate it more deeply, though perhaps only in distant generations.

Frege presumably never read Trede's *Proposals*, but the idea of the construction of a 'necessary language' seems to have impressed him. In his own *Begriffsschrift* he appears to have followed Trendelenburg's advice to combine the Leibnizian program with the Kantian distinction of form and content. In 1882 he wrote:[44]

> I have tried now to supplement the mathematical formula language with signs for logical relations, so that initially a *Begriffsschrift* for the area of mathematics should come out of it. . . . The use of my symbols in other areas is not excluded through this. . . . Whether this happens or not, the intuitive representation of the forms of thought has in any case a significance that goes beyond the area of mathematics. For that reason, may philosophers pay some attention to this matter.

In the essay of which these are the last sentences Frege justifies his *Begriffsschrift* in terms that are mostly taken from Trendelenburg, but without the historical details. He argues that signs are necessary for fixing our fleeting impressions (*ibid.*, p. 107). 'Without signs we would hardly rise to conceptual thinking' (*ibid.*). A concept is reached only by having a sign refer to it; 'since it is in itself non-intuitive, we require an intuitive representative to make it appear to us' (*ibid.*, pp. 107–8). But ordinary language is a deficient medium. 'Language is not determined by the logical laws in such a way that adherence to the grammar guarantees the formal validity of our movements of thought' (*ibid.*, p. 108). Such flaws have their root 'in a certain softness and changeability of language, which on the other hand is the condition of its ability to develop and of its manifold usefulness' (*ibid.*, p. 110). Ordinary language can be compared to a human hand which, in spite of its adaptability, is not sufficient for all purposes. That is why we create artificial hands, tools for specific uses. Their precise function is made possible because of the rigidity of their construction, the inflexibility of their parts (*ibid.*). The arithmetical formula language is the partial realization of a *Begriffsschrift*, but it must be supplemented so that it can express the logical forms (*ibid.*, p. 112). Such a project can be realized only if it is accompanied by a *philosophical* analysis of the concepts to be represented. Frege thinks there is no need to worry about the possibility of such a project since he has shown it is feasible and has already tried to realize it in his new logic.

It is useful to detail the similarities of Trendelenburg's and Frege's expositions because this draws attention not only to the fact *that* Frege had read Trendelenburg, but also to *how* he read him. What is for Trendelenburg the topic of a historical essay becomes for Frege the impetus and justification for the construction of a *Begriffsschrift*. Though he draws liberally on Trendelenburg's essay his reading is selective and problem-oriented. Trendelenburg's Aristotelianism is, obviously, of no interest to him. In the whole body of Frege's writings there is not a single indication that he ever read the logical writings of Aristotle or spent any time investigating Aristotelian logic. He criticizes psychologistic logicians and Booleans, but there is no criticism of the theory of the syllogism. Whereas Trendelenburg eventually rejects the narrow scope of Kantian formal logic in favor of his own (Aristotle-inspired) logic of purpose, Frege completely ignores these conclusions and takes Trendelenburg precisely as confirming him in his own Kantian viewpoint. The explanation of the way Frege reads Trendelenburg is, of course, that he is not concerned with historical understanding but with an interpretation of Trendelenburg's text that connects it with his own theoretical problems.

5 *Lotze's conception of a pure logic*

By the time he wrote the *Begriffsschrift* Frege had also read Lotze.[45] He read him in the same selective, problem-oriented way he read Trendelenburg. There is no indication of any interest in Lotze's metaphysics, his aesthetics, or his philosophy of religion. It is a Lotze stripped to the logical bones that appears in Frege's thought. What he finds in Lotze is what he had found in Trendelenburg: a joining of Leibnizian and Kantian ideas, but more deeply and more systematically thought through. But with respect to his logical and epistemological views Frege's thought is indeed 'not independent of Lotze's,' as Bruno Bauch, Frege's colleague and associate at Jena, put it.[46]

Lotze's *Logik* of 1874, an elaboration of an earlier work of 1843, influenced not only Frege, but all philosophical reflection on logic by the post-naturalists. From the viewpoint of Neo-Kantianism, Bauch wrote in 1918:[47]

> Of everything that has followed in the area of logic from Hegel to the present day, there is nothing that has surpassed Lotze's logical achievements in value. . . . His influence reveals itself in every important figure in the area of logic no matter what philosophical direction he might belong to. If he has any claim to significance in logic, he cannot have remained uninfluenced by Lotze.

And from the somewhat different viewpoint of Husserlian phenomenology, Heidegger a few years earlier had called Lotze's *Logik* 'the fundamental book of modern logic.'[48]

Bauch stresses four important features of Lotze's thought. Since they were singled out by him with Frege's work in mind as one of the later developments against which Lotze's achievements must be measured they deserve our special attention. Bauch stresses (1) Lotze's anti-psychologism, (2) his distinction between an object of knowledge and its recognition, (3) his reformulation of the Platonic theory of ideas as an ontology-free theory of objectivity, and (4) his account of concepts as functions.

From the viewpoint of mid- and late-nineteenth-century thought it is clear that any logic which does not consider itself an integral part of empirical psychology requires a special justification. Lotze therefore begins his investigations with the demand for a sharp separation of psychological and logical questions. By insisting on this separation he was consciously turning back to Kant, who had been the first to denounce psychologism; even though, for obscure reasons, Kant is sometimes considered one of the sources of psychologism in logic. But that is a misunderstanding. According to Kant, logic must be formal and the forms of human thought are not empirical. Hence logic itself, in so far as it is concerned with the investigation of the forms of thought, must be a 'pure' *a priori* science. When applied, it may become dependent on empirical and psychological considerations; but pure logic 'has nothing to do with empirical principles, and does not, as has sometimes been supposed, borrow anything from psychology, which therefore has no influence whatever on the canon of the understanding.'[49] Given the Kantian thesis that 'what reason produces entirely out of itself cannot be concealed' (*ibid.*, A xx), it is natural for him to conclude that logic, once conceived, will immediately turn into a 'closed and completed body of doctrine' (*ibid.*, B viii). For this reason, and not just out of historical ignorance, Kant views sceptically *all* of the post-Aristotelian additions to logic. And he holds in particular that 'if some of the moderns have thought to enlarge it by introducing psychological chapters, . . . this could only arise from their ignorance of the peculiar nature of logical science' (*ibid.*).

In following Kant's ideas, Lotze holds that human mental acts contain perceptions and ideas and these are interwoven 'according to the laws of a mental mechanism, but logic begins only with the conviction that this is not the end of the matter.'[50] In logic we must distinguish between ideas as psychological phenomena and their contents (ideas in the objective sense of the word). Such contents can be assessed in terms of truth and falsity and it is with their interconnections that pure logic is concerned. Lotze had already expressed similar views in his controversy with Czolbe, but they are reaffirmed in the systematic context of his *Logik*.

If we follow Bauch and Heidegger in their assessments of the significance of Lotze's *Logik* for the subsequent development of the discipline, we must recognize that this assessment is based largely on Lotze's contribution to the development of anti-psychologism in late-nineteenth-century Germany. It was from Lotze that both Husserl and Frege took the idea and developed it further. In fact the parallels between Lotze's position and Frege's can be amply illustrated. Thus, Frege writes: 'The connections which define the nature of thinking are peculiarly different from associations of ideas. . . . As the external sign of a thought connection we can use the fact that with respect to it the question of true and false is meaningful. . . . The laws of logic cannot be justified through a psychological investigation' (NS, pp. 189–90).

Lotze's crucial distinction is that between subjective mental states and their objective meaning. Such objectivity, 'in general, does not coincide with the reality that belongs to things' (*Logik*, p. 16). The identified content is not assumed to belong to external reality.

> The common world in which others are to find this content to which we are pointing is, in general, only the world of the thinkable; to it we ascribe only the first trace of an independent subsistence and of an inner regularity which is the same for all thinking beings and which is independent of them. And it is therefore of no concern whether any part of this world of thought refers to anything which has an independent reality apart from thinking minds, or whether its whole content persists only with equal validity in the thinking of the thinkers (*ibid.*, pp. 16–17).

Logic does not deal with the external world, but with the world of objective ideas. And this 'inner' world is large enough 'to contain unknown regions which still remain to be discovered by means of systematic investigations' (*ibid.*, p. 190). It is for this reason that a distinction must be drawn between the content that is grasped by us and the way it is grasped, between the object of thought and our thinking of it.

These considerations about the nature of the objective are also reflected in Frege's writings, but not yet in those of his first period of thought. When he wrote the *Begriffsschrift* it was sufficient for him to consider logic as an objective, *a priori* science; only later, when he began to consider more precisely the nature of numbers, did the notion of objectivity begin to take on special significance for him. And at that point he returned to the considerations of Lotze's *Logik*. Of immediate interest, however, is a connected doctrine concerning the relation of propositions to their constituent concepts. Lotze holds that the notion of objectivity (or validity, as he also says) applies directly to whole propositions, but only indirectly to concepts. 'Of them we can only say that they mean something [*bedeuten etwas*]; but they mean something only because propositions are true of them' (*ibid.*, p. 521). Mistaken (ontological) doctrines of concepts have their origin, Lotze argues, in the separation of the concept from the propositional context.

This doctrine also had its ultimate roots in the thought of Kant. Kant had argued against the theory of ideas (and, thereby, in his own eyes, against any naturalistic theory of knowledge) that judgments are not formed out of previously given constituents, but that they possess an initial transcendental unity out of which we gain concepts by analysis. By the late nineteenth century the doctrine had become a standard argument in anti-naturalistic theories of knowledge. Both Sigwart and Lotze were using it in this way.[51] It had become so popular that Wundt could write of the period: 'It had become the dominating characteristic of logic and has in many respects remained so until today to regard the judgment as the beginning of all logical thinking from which the concept was supposed to originate through analysis.'[52] Through Lotze's influence the doctrine also reached Frege, who expressed it in its most memorable form in the context principle of the *Foundations of Arithmetic*: 'Only in a proposition have the words really a meaning. . . . It is enough if the proposition taken as a whole has a sense; it is this that confers on its parts their content' (F, p. 71). Although in this form the doctrine seems to occur almost exclusively in the *Foundations of Arithmetic*, it can be shown that it was already present at the time of the composition of the *Begriffsschrift*, that it guides that composition, and that the doctrine remains an integral component of Frege's thought throughout his later development.

In his *Logik* Lotze is concerned throughout to show the insufficiency of empiricism. He holds that it needs to be replaced by some form of apriorism. 'It is essentially the view of Kant which I defend here and from which German philosophy should never have departed' (*Logik*, p. 536). The empiricist position cannot be supported by introducing questions concerning the genesis of our knowledge (*ibid.*, p. 524); such questions are irrelevant when considering its justification. This is a theme later revived by Frege. Both he and Lotze accuse empiricism of a genetic fallacy, of trying to substitute a historical account of the origin of some belief for a reasoned justification or proof of it. As Frege says explicitly later on: 'Although each of our judgments is causally determined, not all such causes are justifying reasons. Because an empiricist direction in philosophy fails to consider this distinction, it concludes that all our knowledge is empirical, because of the empirical causation of our thinking' (NS, p. 2). For both Lotze and Frege this observation implies that logic is a completely non-historical science, since historical explanations are always genetic.

There is a second point on which Lotze and Frege agree in their opposition to empiricism and their defence of the possibility of logic as a pure science. Lotze holds that our knowledge cannot have arisen out of a summation of individual experiences. 'Somewhere it is always necessary to presuppose one of those thoughts whose claimed universality one admits with immediate confidence once the content has been thought' (*Logik*, p. 540). In his *Foundations of Arithmetic* Frege takes up this theme by opposing those 'who recognize induction as the sole original process of inference (and even that as a process not actually of inference, but of habituation)' (F, p. xi). He rejects the 'prejudice' that all knowledge is empirical (*ibid.*, p. 9) and asserts that even induction 'can be justified only by means of general propositions of arithmetic' (*ibid.*, pp. 16–17).

Lotze's theory of concepts is best understood against the background of these discussions, though Bauch in his review makes the correct critical observation that Lotze's treatment of the logical theory of concepts is strangely at odds with what he has said about the priority of propositions (or judgments) over concepts. For when he turns to the discussion of concepts he insists that logic must follow the traditional order and account for concepts separately from and prior to judgments.[53] Bauch points out that this gives Lotze's discussion a psychologistic and atomistic appearance that seems incompatible with his other assertions.

According to Lotze, concepts are formed out of simpler ones as marks of the new concepts. But the combination of marks is not always a mere conjunction. 'In general, marks are not co-ordinated into concepts in the same way' (*ibid.*, p. 47). That is why the correct symbol for the construction of a concept is not the equation $S = a + b + c + d + \ldots$, but the designation $S = F(a, b, c, \ldots)$ in which the mathematical expression indicates only that a, b, and c must be combined to form the value of S in a way that can be given precisely in every special case, but which in general is most varied (*ibid.*). In his review, Bauch relates this Lotzean account to both Kant and Frege:

> The notion of a function which was taken by Lotze from mathematics and made fruitful in logic has received a brilliant development in mathematics again on the basis of logic. The altogether classical proof of this is the mathematical work of Frege. The interrelation of logic and mathematics prepared by Lotze also explains the tight connection between Lotze and Kantian philosophy, not least with reference to the notion of function. For it is through Lotze . . . that Kant's idea that transcendental laws or forms, just as much as judgments, are really functions and that concepts rest on functions receives its further elaboration and reformulation (*ibid.*, pp. 47–8).

It is indeed one of Lotze's aims to show how logical and mathematical notions are interconnected, as Bauch points out. And his theory of concepts as functions is part of that enterprise. Though his theory of inference is for the most part concerned with a discussion of the doctrine of the syllogism, he considers it necessary to search for 'new forms of thought' in other areas of science (*ibid.*, p. 131). It is for him an integral part of conceptualization that our concepts contain quantitative aspects (*ibid.*, pp. 31ff). Mathematics is really a part of logic.

> One must not forget that at least calculation also belongs to the logical operations and that it is only the practically justified division of teaching that makes us overlook the fact that mathematics really has its proper home in general logic (*ibid.*, p. 138).

But mathematical truths are not derivable simply from the principle of identity (*ibid.*, p. 583), which Kant had tried to make the exclusive basis of logical truth. Interesting mathematical equations are not mere trivial identity statements, since in them we express the identity of the different. What is needed is an account of identity statements in which we distinguish the form of the expressions on each side of the identity sign from their content. Two such expressions can have different forms and nevertheless possess the same content (*ibid.*, p. 584).

Among the many things that Frege owes to Lotze, the most important is perhaps the idea of logicism. The significance of this idea is not simply that it gives logic priority over mathematics, but also that it reaffirms the fundamental position of logic with respect to all human knowledge. For if it is correct, as the nineteenth century was increasingly coming to see, that science had to go beyond experience through its employment of mathematics, and if logic was indeed more fundamental than mathematics, then pure logic was at the basis of all knowledge.

Frege's interest in logicism went far beyond Lotze's in one important respect. For Lotze logicism was a fundamental truth, but he never undertook to establish it beyond any possible doubt. In fact, he questioned whether this could be done at all. He held that 'the fundamental concepts and principles of mathematics have their systematic place in logic' and that mathematics is 'a branch of general logic' (*ibid.*, p. 34). But at the same time he argued that 'the vast structure of mathematics . . . forbids any attempt to put it back into general logic' (*ibid.*). Frege's project was of course precisely what Lotze had declared impossible. In particular he objected to the idea that one could separate the logical study of the principles of mathematics from the further developments internal to the discipline. Almost as if in reply to Lotze, he wrote: 'If the most general principles and, perhaps, their nearest consequences are assigned to logic, and the further development to arithmetic, that is as if one separated from geometry an independent science of axioms' (KS, p. 103). Logicism remained completely programmatic for Lotze, but for Frege it became the inspiration of the logic of the *Begriffsschrift*, of the reflections of the *Foundations of Arithmetic*, and of the semantic essays of the 1890s.

6 *The influence of Leibniz and Kant on Frege's thought*

One cannot read Frege's writings without becoming aware of the extent to which he feels indebted to the thought of Kant and Leibniz. No other philosophers are mentioned more frequently and none are spoken of with higher respect.

Frege shares with them, first of all, a concern with the power of human reason. Both Leibniz and Kant believe that reason is an irreducible faculty of the human mind, that it can furnish us with knowledge, and that this knowledge is certain and indubitable. While they disagree to some extent about the scope of this power of reason, they are both to be counted as part of the tradition of Western rationalism that began with Pythagoras, Parmenides, and Plato. In a remarkable passage from the concluding pages of the *Foundations of Arithmetic* Frege reveals that he too considers himself part of that tradition. Speaking of his analysis of numbers he says:

> On this view of numbers the charm of work in arithmetic and analysis is, it seems to me, easily accounted for. We might say indeed almost in the well-known words: the real concern of reason is reason itself. In arithmetic we are not concerned with objects which we come to know as something alien from without through the medium of the senses, but with objects given directly to our reason and, as its nearest kin, utterly transparent to it. And yet, or rather for that very reason, these objects are not subjective fantasies. There is nothing more objective than the laws of arithmetic (F, p. 115).

These remarks closely resemble the traditional rhetoric of rationalist thought, but their most direct link is with Kant's comment about the character of the metaphysics of Nature which is envisaged in the *Critique of Pure Reason*. 'In this field nothing can escape us,' Kant writes. 'What reason produces entirely out of itself cannot be concealed, but is brought to light by reason itself immediately the common principle has been discovered' (A xx).

Like Leibniz and Kant before him, Frege is interested in the study of logic and the foundations of mathematics because they allow one to ask in a precise form what can be known through reason alone. None of the three is concerned with the empirical and psychological conditions of our logical and mathematical beliefs, but rather with the question of their objective justification. It is in just this sense that Frege can be called a philosophical epistemologist. The basic issue of his *Foundations of Arithmetic* is the question whether or not arithmetical truths are analytic. He explains the issue in these terms:

> When a proposition is called *a posteriori* or analytic in my sense, this is not a judgment about the conditions, psychological, physiological, and physical, which have made it possible to form the content of the proposition in our consciousness; nor is it a judgment about the way in which some other man has come, perhaps erroneously, to believe it true; rather it is a judgment about the ultimate ground upon which rests the justification for holding it to be true (F, p. 3).

It is this concern for 'the ultimate ground' of truth and the belief that for at least some truths this ultimate ground is to be found in reason itself, that connects Frege with Leibniz and Kant.

The close link between Frege and Leibniz and Kant (and, of course, also Lotze) is clear evidence that he was in no way motivated by the anti-idealistic realism that Dummett has ascribed to him. External circumstances also confirm this conclusion. In 1919 Frege joined the *Deutsche Philosophische Gesellschaft* which had been founded two years earlier by Bruno Bauch and others (cf. WB, p. 9). The goal of the society, according to its own description, was 'the cultivation, deepening, and preservation of German individuality in philosophy in the spirit of the German idealism that was founded by Kant and continued by Fichte.'[54] The *Beiträge*, in which Frege published his papers 'The Thought,' 'Negation,' and 'Thought Connections,' was the official organ of the society. In their editorial comment the editors said in the first issue of the journal that, unlike other publications, theirs was to have a unified purpose. 'The final goal at which we aim is the completion of the scientific view of the world and of life for which the foundations were laid a hundred years ago during the heyday of German idealism' (*ibid.*, p. 2). The idealistic spirit, they said, had created lasting values 'which must not perish in a dull flattening out of all national individuality' (*ibid.*, p. 1).

PHILOSOPHICAL RECONSTRUCTION

The mission of the German idealistic spirit with all its emanations into the general cultural life was not yet finished (*ibid.*).[55]

We need not assume that because Frege joined the society and published in its journal he wholeheartedly and unconditionally subscribed to all its pronouncements. But he certainly cannot have been motivated by strongly anti-idealistic sentiments. What tied him to the idealists was primarily his opposition to the various forms of naturalism of his time. He shared the idealists' anti-psychologism, their belief in an objectivist epistemology, and their apriorism and rationalism. One can read a good deal of Frege without ever raising the question of the extent to which he commits himself to further tenets of transcendentalism. Anti-psychologism, objectivism, and rationalistic apriorism are compatible not only with Leibniz's, Kant's, and Lotze's idealism, but with a Platonic realism as well. But there are several considerations that make such a reading of Frege difficult, if one is concerned with historical accuracy. First, for anyone raised in the environment of Kantian criticism the adoption of Platonic realism would prima facie appear as an unwarranted piece of dogmatic metaphysics. It would therefore require a strong argument to show that Kantian transcendentalism is insufficient to account for the desired anti-psychologism, objectivism, and apriorism. No such argument can be found in Frege's writings. Second, every utterance of Frege's which has been interpreted by analytic philosophers as an expression of Platonic realism can in fact be found to match similar passages in Lotze's idealistic theory of validity. Third, there are strewn through Frege's writings statements that appear irreconcilable with Platonic realism. In particular the central role of the Fregean belief in the primacy of judgments over concepts would seem to be explicable only in the context of a Kantian point of view.

If we consider Frege's thought shaped by the philosophical ideas of Leibniz and Kant, we must not overlook the fact that he draws from both of them only selectively. Of Leibniz he writes:

> Leibniz has strewn such an abundance of intellectual seeds in his writings that hardly anybody can measure up to him in this respect. Some of these seeds came to fruition in his own time and through his co-operation; others were forgotten but later rediscovered and further developed. That justifies the expectation that much in his work that apparently lies dead and buried will one day come back to life. Among these I reckon the idea of a *lingua characterica* (NS, p. 9).

As it turns out, the idea of a characteristic language is the one Leibnizian idea that influenced Frege most deeply. It is Leibniz the logician who appears in Frege's writings—not the metaphysician, psychologist, natural scientist, theologian, or historian. The mystical and speculative associations surrounding Leibniz's reflections on the universal characteristic are completely overlooked by Frege. What is foreshadowed in Frege is the interpretation of Leibniz that Couturat and Russell made famous and which, in the eyes of analytic philosophers, has made Leibniz almost one of their own.

It is in a similarly selective way that he speaks of Kant, for instance in this representative remark of 1885:

> Since in mathematics, more than in any other science, the material subject matter stands in the background and is dominated by thinking, since in it the structure of thoughts is particularly richly and finely developed, this science is particularly suited as a basis for epistemological and logical investigations. Here is a treasure trove which is open for a great deal of exploration. Except by Kant, very little has been done in that direction, since competent mathematical and philosophical thinking and sufficient knowledge in both areas are rarely combined (KS, p. 99).

Again, it is the logician and philosopher of mathematics in Kant who is Frege's concern. We find nothing in his writings of Kant's other philosophical interests.

One might explain the narrowness of focus with which Frege looks at the two philosophers he most respects as a result of his own specific professional background or as peculiar to Frege's individual history. Is such narrowness not due to the fact that he was first and foremost a mathematician? The Neo-Kantians (with their wider conception of the subject) saw Frege in exactly this way. Paul Natorp comments that Frege's philosophical insights are remarkable for a *mathematician*, but that his work, of course, lacks full philosophical scope.[56] What they did not (and could not) see was that Frege anticipated the appearance of a new type of philosopher—the philosopher as professional, specialist, and logician. It is precisely the type that has come to maturation in the analytic tradition.

7 Logical formalism as the root of the analytic tradition

The Fregean or analytic understanding of the task of philosophy is foreshadowed in Kantian formalism. Leibniz, in accord with the older tradition, had still been a philosophical universalist, believing the task of philosophy to be that of integrating the totality of human knowledge. The German idealists, with Schelling and Hegel in the vanguard, tried to re-establish philosophy as a universal discipline. In a completely different way even the mid-century naturalists had preserved that conception. Kant alone had insisted on the separation of form and content, consigning the content of human knowledge once and for all to empirical science and redefining philosophy as the investigation of the *a priori* forms of human understanding.

For Leibniz thought and perception were distinguished only by degrees of clarity and distinctness. Kant, on the other hand, draws a sharp line between sensibility and understanding as two essentially different faculties. Frege follows him by distinguishing between reason as the source of logical truth, experience as the basis of empirical knowledge, and *a priori* intuition as the foundation of synthetic *a priori* knowledge. For Frege, as for Kant, there is a sharp distinction between analytic and synthetic truths; for Leibniz all truths are ultimately analytic and it is only the finite character of the human mind that prevents us from seeing them as such. For Leibniz the characteristic language is to be a universal language, designed to express any clear thought, to test the validity of any inference by mere calculation. For Frege the *Begriffsschrift* is a tool designed for a narrower and more specific purpose, the expression of the laws of formal logic.

In this sense Frege remains a Kantian formalist. This kind of formalism must be distinguished sharply from the mathematical formalism that is one of Frege's major targets of attack. According to this latter conception mathematical symbols are uninterpreted signs which can be operated on by formal rules. Mathematical formulas are like constellations of figures on the chess-board. Logical formalism as advocated by Kant and Frege holds that one can separate the formal features of human thought from its specific contents and that logic (and mathematics) deals only with form. Arithmetic is formal for Frege not because it consists of uninterpreted formulas, but because its formulas apply to everything that is thinkable (KS, p. 103).

The forms of human understanding of which Kant speaks are assumed by him to be identical for all human minds. They are fixed and their investigation is to be carried out not by empirical research, but by an *a priori* investigation of the conditions of the possibility of any human knowledge. Philosophy is identified for him as the discipline whose task is just such an investigation:[57]

> All rational knowledge is either *material* and concerned with some object, or *formal* and concerned solely with the form of understanding and reason itself. . . . Formal philosophy is called logic. . . . All philosophy in so far as it rests on the basis of experience can be called empirical philosophy. If it sets forth its doctrines as dependent entirely on *a priori* principles, it can be called pure philosophy . . .

We may ask

> whether pure philosophy in all its parts does not demand its own special craftsman. Would it not be better for the whole of this learned

industry if those accustomed to purvey, in accordance with the public taste, a mixture of the empirical and the rational . . . were to be warned against carrying on at the same time two jobs that are very different in their techniques, each perhaps requiring a special talent and the combination of both in one person producing mere bunglers?

The separation of formal *a priori* from merely empirical questions is, for Kant, supposed to bring about a sharp demarcation between philosophical and scientific questions and to show where and how philosophy is possible in the face of the empirical sciences. It is also supposed to show how philosophy can itself gain the objectivity and reliability of science. And it is finally supposed to bring out in what sense philosophy, so understood, would be not just one of the sciences, but a foundational enterprise. As Wittgenstein was later to characterize this viewpoint:[58]

> Logic presents an order, in fact, the *a priori* order of the world: that is the order of *possibilities*, which must be common to both world and thought. . . . It is prior to all experience; no empirical cloudiness or uncertainty can be allowed to affect it — It must rather be of the purest crystal. But this crystal does not appear as an abstraction; but as something concrete, indeed, as the most concrete, as it were the *hardest* thing there is.

Though logic has a fundamental role in Kant's thinking, it does not possess the significance it has for Frege or the subsequent analytic tradition. One reason is that logic for Kant is the logic of Aristotle with all its limitations. Kant assumes that logical truths are all trivial and that the formal sciences such as arithmetic and geometry could not be founded on logic alone, but require the recognition of an intuitive *a priori*. Frege allows for the possibility of *a priori* intuitions, but his understanding of logic is different from and more encompassing than Kant's. He says explicitly that 'propositions which extend our knowledge can have analytic judgments for their content' (F, p. 104). For Kant understanding without sensibility is empty. In other words, the understanding alone cannot produce objects, they are given only through sensibility. Only sensibility provides us with the required objects to test the truth of the equations of arithmetic. For Frege, reason has its own objects. And hence the laws of arithmetic can be confirmed by reason alone.

In Frege's larger conception of logic we can clearly recognize the influence of Leibniz. The Leibnizian idea of a new logic that has come to him through Trendelenburg and Lotze is what motivates the construction of his *Begriffsschrift*. Unlike Kant, Frege thinks that the philosophical analysis of arithmetic depends on the construction of an adequate language. He agrees with Leibniz that ordinary language can never be sufficiently formal to express the logical laws and relations with any precision. It is everywhere contaminated by psychological notions. Ordinary language is a historical product, shaped by social, political, and, above all, psychological forces. It is not constructed with the logical ruler (NS, p. 288). The purpose of constructing a logical language is for Frege primarily that of separating the logical from the psychological in language. 'The business of the logician is a continuous fight against the psychological and, in part, against language and grammar,' he writes in the 1880s (*ibid.*, p. 7). And again, later: 'Work in logic is to a large extent a battle against the logical blemishes of language, though language is, none the less, also an indispensable tool' (*ibid.*, p. 272).

III

A Language of Pure Thought

1 Frege's idea of a Begriffsschrift

In 1879 Frege published his *Begriffsschrift*, the product of five years of reflection, in which the ideas received from Trendelenburg and Lotze, Leibniz and Kant were transformed into an entirely new kind of logic. Though small in size, the work is perhaps Frege's greatest and most enduring achievement. His later writings about the foundations of arithmetic and problems of meaning may be more philosophical, but they have remained considerably more controversial than this first work. In fact many of his innovations in logic are so solidly incorporated into the body of contemporary logic that it has become difficult to appreciate fully Frege's originality in conceiving them. The logical system developed in the book differs radically from all earlier forms of logic. Frege had taken almost nothing from the traditional logic of the syllogism and had borrowed only very sparingly from Boolean algebra.

Frege's logic effectively brought to an end the dominance of Aristotelian logic which had been taken for granted in the schools for more than two thousand years. Post-Aristotelian logic begins only with Frege. By terminating the life span of Aristotle's system Frege completed a process that had begun centuries earlier with Galileo's destruction of Aristotelian physics. In the field of logic, it was an epoch-making work.[1] As Bochenski put it, the significance of the book is comparable only to that of Aristotle's *Prior Analytic*, and Frege himself must be considered the most important thinker in the field of mathematical logic.[2]

In the seventeenth century Ramus, Bacon, and Descartes denounced Aristotelian logic as useless, but in spite of their criticism it remained part of the philosophical syllabus and continued to shape the epistemological and metaphysical concepts of philosophers. It seemed to lend support to the metaphysics of substance and accident, it fostered doubts about the reality of relations, and it seemed to make plausible a theory of knowledge that is usually called 'the theory of ideas.' Aristotelian logic was modified and added to, but until the nineteenth century no radical reform had been effected. The first steps towards such a reform were taken in the middle of the century by Boole and others, but in the end their work amounted only to a clarification of the concept of Aristotelian logic with the machinery of mathematics; it did not bring about the radical reconstruction which Frege envisaged and was the first to undertake.

Despite everything that separates his logic from Aristotle's it is nevertheless legitimate to consider both as forms of logic. Both give systematic presentations of valid inferences; both provide accounts of the formal structure of judgments; and both deal with the problem of concept-formation.

They agree, above all, that logic is primarily concerned with the assessment of inference and that the question of the validity of an inference is one of an objective matter of fact. In the *Prior Analytic* Aristotle writes: 'A syllogism is discourse in which, certain things being stated, something other than what is stated follows of necessity from their being so. I mean by this last phrase that they produce the consequence, and by this, that no further term is required from without in order to make the consequence necessary' (24b 18). Logic is understood here as a means of determining the necessary consequences of a statement and therefore of determining its objective content. It is for this purpose also that Frege designed his *Begriffsschrift*. At the beginning of his book he asks what the status of mathematical truths is. 'Now when I turned to the question . . . I first had to determine how far one could go in arithmetic through inferences alone, relying entirely on laws of thought which transcend all particulars. . . . So that nothing intuitive could enter unnoticed, everything had to depend on keeping the chain of inference free of gaps' (BS, p. x).

Frege went beyond Aristotelian logic in two important respects. Having started from mathematical discourse, he soon realized that characteristic inferences in arithmetic could not be analysed satisfactorily in terms of Aristotle's syllogistic principles. Secondly and more fundamentally, he concluded that the conceptual content of a statement is in general only imperfectly represented in everyday language. 'I found the inadequacy of language an obstacle,' he writes. 'This deficiency led me to the idea of the present *Begriffsschrift*. Its first purpose therefore is to help us determine in the most reliable manner the validity of a chain of inference and to point out any presupposition that tries to sneak in unnoticed' (*ibid.*). The symbolic notation used to express conceptual contents with complete precision is to be a 'formula language of pure thought.'

This notation, initially designed for the logical analysis of arithmetical formulas, is to have the wider purpose of providing a means of determining the objective, conceptual content of any statement whatever.

If it is a task of philosophy to break the domination of the word over the human spirit by uncovering deceptions about the relations of concepts which arise almost inevitably from common linguistic usage, and by freeing thought from that with which it is infected only by the nature of the linguistic means of expression, then my *Begriffsschrift* can become a useful tool for the philosopher, if it is further developed for these purposes (*ibid.*, pp. xii–xiii).

These claims amount to the statement of a whole philosophical program. We can summarize it as follows:

(1) Meaningful statements possess an objective conceptual content.
(2) That content is only inadequately represented in ordinary language.
(3) It is possible to design a system of notation in which the conceptual content of any statement can be given an adequate and clear expression.

Implicit in this program is a threefold philosophical methodology. The task of philosophy is seen as *the determination of the objective content* of philosophically interesting statements, a *critique* of their expression in ordinary language, and their *translation* into an adequate language. It is a methodology which the analytic tradition has endeavored to carry out. It has done so by adopting the outline of Frege's program but modifying the details.

If we consider Frege's achievement in hindsight we discover that he thought of his own logic and its applications as more definitive than seems warranted. Like Kant, he assumed that the construction of a satisfactory logic should be a finite and completable task. And he considered his logical analysis of arithmetic as conclusively settling the question of the foundations of the subject (cf. F, p. v). But since Frege there has been an abundance of new logical systems. In a sense, his work was not the completion of anything, but a new historical beginning for logic.

Frege himself thought that he had achieved four major things in the *Begriffsschrift*:

(1) The creation of a symbolic language, a 'language of pure thought' — as he put it in the subtitle of the book — which would allow one to represent the conceptual content of a judgment with complete precision;
(2) An illustration of how deductions could be carried out in this language and how a fully axiomatic treatment could be given to the kind of logic that is today called elementary logic (i.e., propositional and predicate logic);

(3) A definition of the mathematical notion of one object's following another in a series and a proof of a fundamental theorem involving that notion, thereby establishing the usefulness of the new logic for the analysis of arithmetic;

(4) What he regarded as the first step towards establishing the *a priori* character of the laws of arithmetic; a necessary step for the vindication of his Leibnizian and Kantian presuppositions.

If he had expected that others would quickly recognize these achievements, he was wrong. A few responses were positive. Kurt Lasswitz, the scientist and author of early science fiction, recommended the book highly in his review, calling it 'a valuable contribution to the theory of thinking.'[3] Abbe, in a confidential report to the Jena faculty, spoke of the 'very interesting cluster of ideas' in the book, but warned that 'it will probably be understood and appreciated by only a few.'[4] The reaction of philosophers to the new logic was silence, as Frege had feared, while those concerned with symbolic logic in the form of Boolean algebra failed to see why Frege had found it necessary to make an altogether new start. Foremost among them was Ernst Schröder, Germany's leading expert in the field of Boolean algebra.

Schröder's review was the most detailed the *Begriffsschrift* received. It tore the book apart—in the politest possible way. There was a good deal of hurt vanity in Schröder's words, even in his very first sentence: 'This very unusual book—obviously the work of an ambitious thinker with a purely scientific turn of mind—pursues a course to which the reviewer is naturally highly sympathetic, since he himself has made similar investigations.'[5] The sting at the end of the sentence could not be overlooked. Why had Frege failed to mention the achievements of Boolean algebra? Why had he failed to appreciate Schröder's own contributions to the subject? The reviewer could not help making himself even more explicit:

> The present little book makes an advance which I should consider very creditable, if a large part of what it attempts had not already been accomplished by someone else. . . . I consider it a shortcoming that the book is presented in too isolated a manner and not only seeks no serious connection with achievements that have been made in essentially similar directions (namely those of Boole), but disregards them entirely (*ibid.*, p. 220).

Schröder noted that Frege had expressly rejected as an artificial similarity 'the interpretation of the concept as a sum of its marks' (BS, p. x). And he well understood that this had been meant as a dig against Boolean algebras.

He therefore proceeded to the counter-attack. Frege's notation, he commented, was inferior to Boole's, since it was more cumbersome and used unfamiliar symbols to express logical relations, whereas Boole had employed the well-known symbols of algebra. 'With regard to its major content the *Begriffsschrift* could actually be considered a *transcription* of the Boolean formula language' (*Conceptual Notation*, p. 221). In other words, the new notation had added nothing to the insights of Boole. In addition, Frege's exposition was repetitious and concerned with details hardly worth expression — details that easily showed themselves as evident 'through mental cross-multiplication' (*ibid.*, p. 229). Admittedly, Boole's algebra had given a somewhat incomplete account of general propositions, and specifically of particular and existential propositions. Frege's account was more precise, but the flaw in Boolean algebra could easily be removed by adding a few special letters to the notation. Frege's attempt to apply his logic to the analysis of the notion of a mathematical series appeared 'very abstruse — the schemata are ornate with symbols' (*ibid.*, p. 230). Besides, it was difficult to see the interest in this undertaking since Frege's ultimate project, the logical analysis of arithmetic, had already been achieved in large part 'through the perceptive investigations of Hermann Grassmann' (*ibid.*, p. 231). Having done his job, Schröder let it be known that his overall purpose was to encourage the author of *Begriffsschrift* to further research, rather than to discourage him.

Frege's silence about Boolean algebra had, of course, been quite deliberate. He saw his own undertaking not as a continuation of Boole's work, but as a new beginning — both more general and deeper than anything Boole had done. Schröder's review convinced him that the purpose he had set himself in the *Begriffsschrift* had not been well understood. In response to Schröder's challenge, Frege therefore set out to show that his logic could easily stand up to Boolean algebra. When he sent off the completed study, entitled 'Boole's Calculating Logic and the *Begriffsschrift*,' he found that neither the editors of mathematical journals nor the editors of philosophical ones were willing to publish it. In the end it remained unprinted until it appeared in the *Nachlass*.

The interest of the essay lies in the fact that it was written almost immediately after the *Begriffsschrift* itself and that there is no other place where Frege explains the motivations for his new logic in equal detail. In it he shows by examples that his notation is capable of expressing complex mathematical propositions (NS, pp. 23–9); he gives a formal derivation of an arithmetical proposition (*ibid.*, pp. 30–6); he examines a logical problem discussed by Boole and Schröder and illustrates how it can be dealt with easily in his own logic (*ibid.*, pp. 45–51). But it is not these things that throw most light on his conception of logic. Of greater interest are his comments about the nature of his symbolic notation and the views he expresses about concept formation, to which he ascribes his account of general propositions. He considers himself at variance with Boolean conceptions in both respects.

As Frege understood him, Boole had been concerned with a limited, practical task: the design of a *technique* that would allow the systematic resolution of certain logical exercises (*ibid.*, p. 13). Boole had taken well-known algebraic symbols and given them various logical interpretations. He read algebraic letters as standing for classes and interpreted them as standing for propositions. And he had shown how one could resolve problems of both class logic and propositional logic by mechanical calculating procedures performed on the algebraic symbols. Jevons had even invented a machine for this purpose (*ibid.*, p. 39).

Frege did not want to dispute the significance of such technical improvements in logic. On the contrary, he conceived of his own system also as comparable to a technological advance — as, indeed, itself a technological advance. His new symbolic notation, he said, was related to natural language as a mechanical hand is to a human hand (BS, p. 110) or as the microscope is to the human eye (*ibid.*, p. xi). Theoretical conception and technical execution must go hand in hand. When, in another context, he discussed the correspondence between Leibniz and Papin, he drew special attention to the fact that Papin could never carry out his plans for a steam engine and Leibniz could never complete his plans for a calculating machine because of 'the imperfection of tools' (*ibid.*, p. 96). Much time had to elapse between the first conception of the idea and its full realization.

> I believe that the steam engine would have reached a high degree of perfection already during Papin's time if our present technical resources had been available. But of course it needed the stimulus which the idea of a steam engine gave to create these tools. . . . With respect to his calculating machine Leibniz considered himself lucky to have seen its rough realization. The difficulties were presumably similar to those in the construction of the steam engine and lay in the imprecision of labor at that time (*ibid.*).

Although such technical improvements were important, the purpose of the *Begriffsschrift* was not restricted to them. 'The proper field of observation for logic lies in the scientific work-place' (NS, p. 37). The new logic was supposed to provide a tool for the analysis of arithmetic. 'Apart from a few formulas which were mentioned because of the Aristotelian forms of inference, only those were included which seemed necessary for the proof' of the mathematical theorems of the third chapter of the *Begriffsschrift*. Because of the specific problem Frege set himself, he could not employ algebraic symbols to express logical relations, as Boole had done. 'Someone who demands that the relations of signs should be as far as possible in accord with the facts will always consider it a perversion of the real situation for logic to borrow its symbols from arithmetic. The subject matter of logic is correct thinking and that means the foundation of arithmetic as well' (*ibid.*, p. 13). His logic was to be a *lingua characterica* suitable for the expression of actual mathematical contents and not, like Boole's, a mere *calculus* restricted to pure logic.

2 Logical symbolism and language

What had irked Schröder about Frege's notation, and also bewildered more sympathetic readers (like Peano and Russell), was of course the strange appearance of Frege's symbolism. *Functionally* that symbolism corresponds to the contemporary notations for propositional and predicate logic, but in its external shape it is quite different. Both the Boolean notation and that of contemporary elementary logic (which derives from Peano and Russell) adapt themselves to the ordinary way of writing sentences in natural language, that is, they retain the linear, sequential character of such writing. Frege had tried to rethink the problem of notation in a more radical way. He reasoned that the linear appearance of ordinary writing was due entirely to the fact that a written sentence of natural language represents a spoken sound which possesses a linear temporal order. But the *Begriffschrift* was not to be bound by the requirements and conventions of ordinary language. His symbolism was to be a written language, not a spoken one. Hence he felt justified in making use of the two-dimensionality of the page in a way which our common written language does not (cf. BS, p. 111). His symbolic notation therefore has a strangely geometrical appearance—as Schröder was the first to notice. It looks like a series of diagrams rather than an arrangement of propositions and formulas. In Frege's symbolism the various propositions (or propositional forms) which are tied together by connectives are displayed separately, each on a line of its own, with the logical connectives forming a network of lines to the left of the sequence.

Once one gets used to the notation, one discovers in it a clarity which contemporary notations lack. Nevertheless, Frege's departure from common forms of writing proved too radical to find acceptance and the peculiar appearance of his symbolism no doubt contributed to his difficulties in getting his writings published and understood. In addition, it seems from a contemporary point of view that Frege had underestimated the usefulness of a linear notation. His symbolism may possess greater clarity, but it does not lend itself easily to the kind of meta-theoretical study of symbolic systems which has become a major concern of logicians since the work of Hilbert, Gödel, and Tarski.

What was at stake in Frege's rejection of Boole's algebraic notation was more than a quarrel over signs. In giving algebraic equations a logical interpretation Boole had implicitly expressed a certain view about the relations between logic and mathematics. Frege agreed with Boole that logic and mathematics needed to be brought into closer contact, but he disagreed on the question of which of the two sciences was more fundamental. Boolean algebra had been an attempt to reform the old syllogistic logic that many besides Frege had found deficient. Such attempted reforms had almost all been the work of mathematicians (Boole, Hamilton, Jevons, Schröder) and it was natural for them to conceive of the reform as a process of introducing mathematical concepts and ideas into logic. The whole idea of a formal notation for logic had been taken over from mathematics. Leibniz himself had justified his search for a logical language by emphasizing the importance of an adequate notation in algebra and, even more, in calculus.

Frege was not in total disagreement with these ideas. He had called his *Begriffsschrift* a language 'modelled upon the formula language of arithmetic' (BS, p. vii). And he had stated as his goal the achievement of the same kind of precision in logic which Euclid had achieved in geometry and, indeed, a further improvement on Euclid's axiomatic method. What was important for him was the ability to express in the logical language chains of inference which were free of gaps and did not require recourse to intuition (*ibid.*, p. x). But he required such mathematical rigor in order to show that arithmetic was only an extended logic or a proper part of logic. 'There arises here the task,' he wrote, 'to set up signs for logical relations in such a way that they are suitable to melt together with the mathematical formula language so that at least for one area they form a complete *Begriffsschrift*. This is the point from which my little pamphlet takes off' (NS, pp. 14–15).

The Booleans, on the other hand, tended to think of logic as a particular kind of algebra and hence as a proper part of mathematics. Such reductionism was repudiated by most philosophers who understood the issue at the time. How could logic, which was the basis of all reasoning, be only a part of the specific form of reasoning that is called mathematics? At the time when Frege was battling with Boole and Schröder, Lotze brought out the second edition of his *Logik*, in which he added a special note on the calculus developed by Boole and Schröder. He granted that a formal calculus might allow the 'mechanical resolution of certain exercises,' but he suggested that the rules for such a calculus would have to be constructed 'on the basis of purely logical principles alone and without risky and unclear analogies to the area of magnitudes.'[6] He complained that some of the formulas of Boolean algebra and some of its calculations have no coherent logical interpretation. 'The validity of a logical law is not determined by the structure of the formula, on the contrary the symbolic usefulness of the formula is determined by its correspondence to the meaning of the logical law' (*ibid.*, p. 259). The Booleans did not fail to recognize the difference between arithmetical and logical calculation, but they entertained the idea of an even more general mathematical algorithm for which this difference in possible applications would be unimportant. Against this idea Lotze held that 'the specifically logical laws must stand on their own feet, and it is both an incorrect and a confused thought to look for their justification in some more abstract mathematics' (*ibid.*, p. 260).

Given that Lotze and Frege both subscribed to the reducibility of arithmetic to logic, it is not surprising to find that they also agreed in the rejection of the doctrine that logic is a proper part of mathematics. The Boolean view was, of course, a product of the growth of mathematics during the nineteenth century, whereas Lotze's and Frege's position represented a more traditional and more philosophical viewpoint. In the twentieth century philosophers in general and logicists in particular have tended to uphold the priority of logic over mathematics, while mathematicians in general and in particular those committed to the doctrines of intuitionism and formalism have tended to the opposite view.

If Lotze and Frege could agree on the question of the relation between logic and mathematics, there were other issues on which they did not see eye to eye. Lotze was wary not just of Boole's attempt to mathematize logic, but of the whole idea of formalization. Undertakings such as Boole's, he said at the end of his note, have often announced the beginning of a new epoch in logic, but if the old logic was ever forgotten it would eventually be rediscovered and hailed as the true one. Lotze readily acknowledged the significance of language for logic. As early as 1843 he had written: 'Language is much closer to logic than is generally supposed.'[7] He also understood that our common language reflects the logical facts only imperfectly and incompletely. 'The structure and use of language does not at all coincide with the achievements of thinking.'[8] But that insight had not led him to an interest in the Leibnizian project of a logical language. He was suspicious of those who would substitute formulas for thought or calculation for reflection. In the final sentence of his *Logik* he expressed the hope that German philosophy would once again rise 'to understand the course of the world, and not merely to calculate it' (*ibid.*, p. 608).

To these remarks Schröder was later to reply, in defence of formalization, that if we could only calculate the course of the world we would certainly understand it as far as understanding is at all attainable in this world. Nevertheless, Lotze's doubts reflect an antipathy of much of German philosophy to the new formalized logic, whether in the Boolean or the Fregean form — an antipathy that lasted till deep into the twentieth century. In part that antipathy resulted from a feeling of insecurity in the face of the unusual and hermetic appearance of the new logic. In part it was fear of the incursion of mathematics into an area that traditionally had been integral to philosophy. In some cases, however, the rejection of formal logic was based on more explicit reasoning.

In those places in which the radical empiricism of the scientific naturalists had its effect, formalization, whether in empirical science or in logic, was looked upon with great suspicion. That is true of the criticial positivists, for instance. They completely rejected the idea that language could be used to represent things as they really are, and so they were committed *a fortiori* to the belief that a formal language could not do so. Such a language might have certain practical uses, but it could not show the *real* logical structure of our thoughts. When Fritz Mauthner came to reflect on the disagreement between Lotze and Schröder about the relative merits of ordinary language versus a formal symbolism, he agreed with neither view. In what did their conflict really consist, he asked:[9]

> Only in this, that Lotze considers the abstract words of philosophy as most suitable for explaining to himself and others the course of the world whereas Schröder considers the mathematical symbols which lie outside common language most suitable. Lotze really says: the concepts of philosophical language are clearer than mathematical concepts; it is impossible with mathematical abstractions to get beyond the insights that are reachable through language. And Schröder really answers: the mathematical abstractions are clearer than the abstractions of language.

To which Mauthner adds:

> Clearer, yes, because they are formal and as long as they remain formal; but really empty because human language is not mathematical. . . . It is the merit of O. F. Gruppe to have first explicitly pointed out this difference between logic and mathematics.

For Mauthner himself language is not an abstract system; it is 'only a process, an infinitely complex activity, not a tangible, material organism.'[10] It exists only in the interaction of groups of human beings. 'Every human being is a small virtuoso in his native tongue for whom the complex movements of his organs of speech have become reflexive movements or instinctual actions, if you wish.'[11] The complex social interactions made possible by such movements depend on a constant process of habituation, on borrowing and imitation. 'Language has grown like any big city: room by room . . . house by house, street by street . . . and all this is boxed together, tied together, smeared together.'[12] We like to imagine that thought, logic, and grammar rule our language. 'I would regard it as the proudest result of my investigation, if I could convince people of the unreality, the worthlessness of this trinity of goddesses, for the service of non-existent gods is always a heavy burden, hence always noxious' (*ibid.*, p. 11). No appeal is possible to reason or logic outside language. 'Language . . . exactly represents reason. . . . A people knows no other reason, no other logic than that of its language' (*Die Sprache*, p. 86).

Because language exists only as a historical and social process, there cannot be any artificially invented language. It is not that science is barred from inventing more convenient or more rational notations. But these can never lay claim to being real languages; they can never replace natural language, since they are rooted in it and are mere outgrowths of it (*ibid.*, pp. 35–6). Mauthner, unsympathetically, thinks of all artificial languages as similar to Esperanto or Volapük. In his eyes Boolean algebra, symbolic logic, and these new languages were all products of the same mentality. It was typically mathematical logicians like Couturat[13] and chemists like Ostwald,[14] scientists who worked daily with artificial symbols, who were also the prophets of Esperanto as a new universal language. Mauthner, on the other hand, had no taste for such an idea. For him, Esperanto was no language at all, 'because a language can only grow between human beings, it cannot be invented by a single human being' (*Die Sprache*, p. 42). There could be no ideal language:

1. because the catalogue of the world which would have to stand at the basis of such an ideal language has never been provided by our natural sciences and can never be provided.
2. . . . because the artificial language would again be no ideal one, but only an artificial, juvenile translation of a given language, e.g., the native language of its inventor (*ibid.*, p. 33).

Mauthner's view of language was thoroughly naturalistic and for that reason close to psychologism. 'This psychologism would be the truth, if our psyche did not have to speak in order to make itself understood' (*Wörterbuch*, vol. 3, p. 169). Individual psychology has discovered that there is no psyche and hence has had to turn physiological; in social psychology we must consider language the *sensorium commune* of the social group (*Die Sprache*, p. 28). Mauthner's writings on language are always scholarly and full of surprising insights, they are often witty and sometimes profound, but it also seems that his naturalism prevents him from fully understanding the significance of formal symbolisms as model languages and as models of language. Wilhelm Jerusalem, who was as close to Mach's naturalism as Mauthner, perhaps first saw this blind spot in Mauthner's thought. He felt Mauthner's critique of language was of great significance to scientific research. 'But since logic cannot do without language, the attempts to create and introduce a kind of universal conceptual language that have been made again and again since Leibniz are also of great logical significance.'[15]

Mauthner's conclusions about the powers of language were more sceptical:

It is not the case that something becomes common because all consider it good; but rather what accidentally has become common, all afterwards regard as good. And as sacred at that. It is hard to say how far this instinct has already taken poor mankind and how far it will still take it. This grotesque conclusion throws sharply illuminating light even into the area of science; even the concepts or signs of knowledge become valuable and lead a stretch further only if they are common to the experts in the field; for the concepts or signs become useful only as common symbols; for a time we work together, until the crookedness or falseness of the signs has become apparent, then there follows a period of revolution, and once again we hurry to agree on new common symbols (*Die Sprache*, p. 90).

From such sceptical relativism no road leads back to the Leibnizian idea of a logically perfect language. For Mauthner those who search for such a language are mere 'clever nose-pickers who want to invent a new system of formal logic, who try their wits on symbolic logic or algebra of logic, who dream of a precision tool of thinking, a new thinking-machine, and who with all their admirable mathematical clarity none the less fall back into the error of formal logic' (*Wörterbuch*, vol. 1, p. 446).

Frege, Schröder, Lotze, and Mauthner present us with four roughly contemporary views of the interrelations among language, logic, and mathematics. While Frege and Schröder both believe in the need for formalization, Lotze and Mauthner both reject the idea. While Frege and Lotze both assert the primacy of logic over mathematics and therefore the reducibility of mathematics to something more fundamental, Schröder, in the Boolean tradition, maintains that mathematics is more fundamental than logic and Mauthner, following Gruppe, separates mathematics and logic altogether. While Frege and Mauthner both hold that logic is closely related to language, Schröder conceives of his logical notation as a kind of calculus and Lotze assumes that logic is concerned with conceptual understanding that transcends both language and the mathematical calculus. The historical meaning of Frege's conception is defined by the place it occupies among these alternative possibilities.

3 Frege's propositional logic

The first and perhaps most important step in the construction of Frege's new logic is his analysis of an actual judgment into two basic components: the judgeable content and the judging of that content. Logic for Frege is not concerned with how and why we judge but only with the properties of the conceptual contents of judgments and their interrelations. This distinction is what guarantees the separation of logic from psychology.

In ordinary language the separation is obscured by the lack of a separate sign that would indicate whether a given sentence is asserted or not. For that reason it is often said that the predicate of the sentence has 'assertive power'; but assertion and predication must be strictly separated. Later on Frege writes:

> One can express a thought without asserting it. But in ordinary language there exists no word or sign whose exclusive job is to assert. That is why even logic books confuse predicating with judging.... To think is to grasp thoughts. After a thought has been grasped, one can recognize it as true (judging) and express that recognition (asserting) (NS, p. 201).

This is an often-repeated theme in Frege's writings.

In the late essay 'Negation' Frege writes: 'The linguistic usage of everyday life is best hit upon if one understands by a judgment an act of judging, just as a jump is an act of jumping' (KS, p. 370). In contrast to a prevailing tradition he holds that there is only one such act which is of interest to logic. Judging is always recognizing something as true (*ibid.*, p. 149). The tradition to which he opposes himself with this claim holds that there are two forms of judgment, assertion and denial. In Frege's view that conception involves avoidable technical complications. 'If we can do with one form of judging, then we must do so' (*ibid.*, p. 374). The opposite view is not only technically disadvantageous, but involves a false conception of what judging really is. It derives from the belief 'that the judging person establishes a connection and order in the parts through his act of judging and that the judgment is brought about in this way' (*ibid.*, p. 371). If we assume that the person who judges establishes a link and order between the parts of the judgment, it seems also reasonable to suppose that he has the power to destroy that link. 'Thus judging and denying appear as a pair of opposite poles which, as a pair, are of equal rank' (*ibid.*). But judging involves first of all the grasping of a complete thought, and then the recognition of its truth. The thought is not created through the act of judgment. 'Even grasping a thought is not creating a thought, nor is it the establishing of an order among its parts' (*ibid.*). Even when we deny something we grasp a thought in which the parts stand in a definite order, we do not destroy the thought.

In its form in the *Begriffsschrift* the doctrine is derived directly from Lotze's *Logik* (p. 61), where we read:

> We imagine a particular relation (of whatever kind) between S and P expressed as a thought still open to question through the judgment 'S is P'; this relation expresses the thought content, about which two opposed subsidiary judgments are made; the one, affirmative, gives it the predicate of validity or reality, the other, negative, denies it....

But this difference does not give rise to two essentially distinct forms of judgment. Validity and invalidity are rather to be considered in this connection as factual predicates which hold the total content of the judgment as their subject.

Frege's primary logic is the logic of unanalysed judgeable contents. Such contents are either affirmed or denied, he says in the *Begriffsschrift*. The terminology may be slightly misleading at first, since it seems to revert to the confusion between a judgment and its judgeable content. What, after all, is the difference between saying that a judgment is asserted and that a judgeable content is affirmed? Moreover, the distinction of affirmation and denial seems to be in direct conflict with Frege's claim that there is only one act of judging, namely, judging something to be true, and not the two acts the traditional textbooks had spoken of (cf. BS, pp. 1–4). Closer reading shows that Frege uses the terms 'affirmed' and 'denied' in the *Begriffsschrift* in the sense of 'true' and 'false.' In his later writings the terminology is changed accordingly. That leaves the question why he should have adopted the terminology of the *Begriffsschrift*, which sounds more subjective and psychologistic than Frege's views warrant. The most likely explanation is that the terminology is a residue of an earlier stage of Frege's development in which the distinction between a judgment and its content was not yet completely clear to him.

With the distinction between affirmed and denied judgeable contents in hand Frege can proceed to characterize the fundamental notions of his propositional logic. There are two of them, negation and the conditional. A judgeable content which is negated is thereby transformed into another judgeable content which is denied when the first content is affirmed and affirmed when the first is denied. The conditional is introduced analogously. By combining two judgeable contents A and B with a conditional sign we form a new judgeable content. That content is denied when the antecedent A is affirmed and the consequent B is denied. It is affirmed when A is denied or B affirmed. Frege's conditional is therefore the material conditional of later logic.

Frege treats the conditional he has introduced as a formal analogue of the 'if-then' of ordinary language, but he realizes that the two do not completely correspond. In the late essay 'Thought Connection' he writes: 'It is not my concern in the explanation to hit upon the linguistic usage of everyday life which is too imprecise and unsteady for the purposes of logic' (KS, p. 388). The words 'if-then' have a multitude of functions in ordinary language. For instance, a speaker can use them to give a reason why something is true, or he can express a supposed causal relation by means of them. For any uttered sentence, we must distinguish between the 'logical kernel' of the utterance and the hints and suggestions a speaker might give by uttering the sentence. In explaining the characterization of the conditional, Frege says: 'My task here is to uncover the logical kernel of a connection of two thoughts by separating out the inessential accessories' (*ibid.*).

When reconsidering his truth-functional characterization of the conditional, he writes in 1906:

> It is now twenty-eight years since I formulated this explanation. At the time I thought that I only needed to touch upon the point and others would immediately know more than I. And now, when more than a quarter-century has passed, the vast majority of mathematicians still have no understanding of the matter, and it is likely to be the same among logicians. What dullness of mind! (NS, p. 202).

They show they do not understand the explanation because they still complain about the lack of an inner connection between the thoughts.

> Let someone give an explanation in which more of the thought is introduced. Whatever one takes from the thought in addition will probably be quite superfluous. The whole matter simply becomes more complex without any gain at all. Or it will turn out that the sentences (antecedent and consequent) are incomplete sentences, so that in fact one has not brought thoughts into connection but concepts or relations (*ibid.*, p. 203).

The purpose of the symbolic notation is thus not to imitate natural language, but, with a minimum of machinery, to allow everything to be said that can be said in ordinary language. 'In place of ordinary language with its proliferating, indefinite logical forms, I put a small number of forms. This seems to me essential for making the movements of thought precise,' Frege writes in 1880 (*ibid.*, p. 44). Explanation always consists of condensing and simplifying. What is needed are signs whose meaning is simpler than those of ordinary language. 'But a content is simpler to the extent to which it says less' (*ibid.*, p. 40). The failure of the material conditional to reproduce all the complexities of the 'if-then' of ordinary language therefore is no failure at all, but is deliberate and necessary. The significance of the symbolism lies in the fact that 'a few signs suffice to express a large number of mathematical relations' (*ibid.*, p. 30).

Let us assume that we can assert a conditional statement consisting of A and B, but that we do not know whether A and B are affirmed or denied. In that case our conditional will correspond closely to the 'if-then' of ordinary language (BS, p. 6). Let us assume, moreover, that we are expressing a *general* relation between judgeable contents by means of the conditional sign; if that generality represents a natural law we would be expressing a causal relation (*ibid.*, p. 23). Our symbolism thus allows us to express all the things we can express by the 'if-then' of ordinary language, but it separates the various ingredients that go into the use of the words 'if-then.' Our symbolism is useful precisely because it is more analytic than ordinary language.

The propositional logic which Frege constructs by means of the operations of negation and conditional corresponds in power exactly to Boole's logic of secondary propositions. That was of course what Schröder had noticed when he called the *Begriffsschrift* a transcription of the Boolean formula language. In fact it is even plausible to assume that Frege modelled his propositional logic on that of Boole. Both assume naturally that propositions (or judgeable contents) are either true or false and that they cannot be both at once. Both consider the truth and falsity of a complex proposition to be determined by the truth and falsity of the simple propositional components. For both, in short, propositional logic is two-valued and truth-functional.

There are nevertheless important differences in the ways they develop their respective propositional logics—differences that Schröder did not fully comprehend. For Frege propositional logic is primary and fundamental; everything else has to be built upon that foundation; for Boole propositional logic is a logic of secondary propositions built on a logic of primary propositions which is a logic of classes and the logic of classes is itself a part of general algebra. Given our ordinary algebraic equation, we can interpret the letters usually taken to indicate numbers as standing for classes, we can interpret addition and multiplication respectively as the logical operation of product and sum, we can take the numerals '1' and '0' to stand for the universe of discourse and the empty class respectively. In this way, algebra becomes a class logic. Class logic can in turn become a propositional logic if we take the letters to stand for classes of moments at which propositions are true and the numerals '1' and '0' as representing the class of all moments and the class of no moments respectively.

When comparing his own logic to Boole's Frege wrote: 'In contrast to Boole I reduce the primary propositions to the secondary. . . . Thus, I believe I have produced in an easy and appropriate manner an organic link between the two parts' (NS, p. 19). Where Boole had treated the two parts of his logic as two separate interpretations of the same algebra, Frege's aim was to 'produce the whole in one piece' (*ibid.*, p. 15). Only in that way could one hope to make logic useful and, in particular, useful for laying the foundations of arithmetic.

Boole's procedure was the result of his unspoken conviction that mathematics is more fundamental than logic. This had determined not only his general approach to logic, but also the particulars of its execution. In propositional logic he had chosen negation, conjunction, and disjunction as the basic logical operations, since they corresponded most closely to the operations of algebra. Frege, on the other hand, preferred negation and conditional as basic operations, since the conditional is directly related to and can be used to express inference relations and these are fundamental to logic. Boole was concerned with logical calculation and had designed his notation to facilitate it. Frege, on the other hand, was interested in inference and derivation. His logic had an axiomatic form rather than an algebraic form.

Frege thought he could recognize another motive in Boole's belief in the primacy of class logic over propositional logic. Since Aristotle the logical tradition had held that the science had to proceed from concepts to judgments and from judgments to inferences. The procedure seemed a natural one, progressing inductively from the simpler to the more complex. Ever since Aristotle's writings were arranged in that order in the *Organon* it had been taken for granted by logicians. Boole's preference for class logic over propositional logic was merely the latest expression of that aggregative point of view.

It was Frege who reversed the order and began his *Begriffsschrift* with the logic of propositions. Later textbooks have followed the same procedure. In Frege's case the motivation was not only the discovery that his order of exposition is simpler and more practical. He was convinced that it reflected the true nature of things. Logic, as he saw it, was primarily a theory of inference and the whole aggregative point of view of the logical tradition was fundamentally misconceived.

There is no equivalent in Boole's logic for the axiomatic treatment Frege gave to his logic in the *Begriffsschrift*. In preferring an axiomatic exposition he saw himself as reviving the Euclidean tradition (cf. BS, pp. ix–x; F, p. 1; NS, pp. 220–1). This method of exposition seemed to him essential if one's purpose was to test the epistemological foundations of mathematics. Inferences had to be constructed in the most rigorous manner; inference lines had to be absolutely clear (cf. BS, pp. 87–8). One had to establish beyond any possible doubt which mathematical propositions could be proved and which could not.

Frege's aim was not simply to apply the Euclidean method, but to improve on it. Euclid had derived all his propositions from a few unprovable assumptions, but he had failed to state the rules of derivation which are needed to justify every step of the deduction. Frege's logic is explicit about the distinction between axioms and rules. He explicitly states as his only rule in the *Begriffsschrift* that of *modus ponens*. Later he realized that substitutions also needed to be justified by a rule and he added substitution rules in the *Grundgesetze*. Frege's logic thus reaches a degree of precision which it took other logicians much trouble to achieve. Speaking of Russell and Whitehead's *Principia Mathematica*, Kurt Gödel could later say:[16]

> It is to be regretted that this first comprehensive and thoroughgoing presentation of mathematical logic and the derivation of Mathematics

from it is so greatly lacking in formal precision in the foundations . . . that it presents in this respect a considerable step backwards as compared with Frege.

The formal precision of Frege's propositional logic has hardly been improved upon. Not only did he give the first axiomatic symbolic formulation of that logic and thereby rescue it from the limbo in which it had existed since the decline of medieval logic, but his axiomatization also proved to be complete. All the valid principles of two-valued propositional logic are derivable from his six axioms.[17] If there is a flaw in Frege's axiomatization it is the rather negligible fact that his third axiom is derivable from the others.[18]

4 *The rejection of subject and predicate*

The details of Frege's axiomatic treatment of propositional logic are of no concern here. However important and interesting that treatment may be, it is not the logically and philosophically most important part of the *Begriffsschrift*. That is to be found in the analysis of general propositions, the introduction of a notation corresponding to the modern quantifiers, and the new theory of judgment which had led him to this analysis. Frege himself later called his notation for generality 'one of the most important ingredients of my *Begriffsschrift*' (BS, p. 105). By its means he succeeded in formulating what amounts to modern predicate logic—a theory which incorporates the traditional theory of the syllogism, but is more extensive, more rigorous, and more useful. With this new notation he could analyse the mathematical notion of following in a series and make plausible the suggestion of Lotze that arithmetic was nothing but an extended logic.

At one point Leibniz had argued that arithmetic consisted of truths of reason, since clearly one could prove the equation '$2 + 2 = 4$' from the definitions '$1 + 1 = 2$,' '$2 + 1 = 3$,' and '$3 + 1 = 4$.'[19] In the *Foundations of Arithmetic* Frege points out that Leibniz's derivation is incomplete, since it relies implicitly on the law of the associativity of addition: '$a + (b + c) = (a + b) + c$.' The analyticity of this principle had not been shown by Leibniz. That was no accidental failure, for Leibniz had no means of showing the analyticity of such general principles. Nor had he shown that arithmetic consists of truths of reason alone, since arithmetic contains both existential and general propositions and his logic provided no adequate account of quantification. Here then was the source of Leibniz's failure—his confinement within the syllogistic conception of logic and its subject–predicate account of judgments. And because Leibniz had failed to establish that arithmetic consisted of analytic propositions, others such as Kant had concluded that such a proof was impossible.

A LANGUAGE OF PURE THOUGHT

To understand Frege's progress over the tradition one must compare his theory of judgment with the traditional one. It had traditionally been held that every judgment is, basically, composed of a subject and a predicate, as in the sentence 'Socrates is a man.' This doctrine was taken to have both metaphysical and epistemological implications. In particular, it was taken to apply as well to general propositions like 'Every man is mortal' and 'Some men are Greek' with 'every man' and 'some men' being considered to be subjects and 'mortal' and 'Greek' predicates of the two sentences. The analysis, which had its origin in the writings of Aristotle, seemed to work well enough in simple cases. Its limit was reached, however, in sentences of multiple generality, such as 'Every son is the child of some father' or 'There is some father who is the husband of some mother such that every one of her children is one of his.'[20] Sentences with multiple generality are perhaps not common in everyday discourse, but they occur in abundance in arithmetic and in theoretical statements of empirical science. It was therefore clear that, as long as no satisfactory account of multiply general sentences had been discovered, no satisfactory logical analysis of the laws of arithmetic was possible.

Boolean algebra had not gone beyond traditional logic in this respect. Even Schröder in his review of the *Begriffsschrift* had to admit the point. But he thought the blemish could be easily fixed up. In Frege's eyes the differences between Boolean logic and his own were more radical. There was a difference in the scope of the two logics which, 'inessential as it may at first appear, depends on the fundamental design of the two' (NS, p. 23). Boole's conception of concepts and their relation to judgment was basically the traditional Aristotelian conception (*ibid.*, p. 16). A new theory of concepts was needed and that meant a new theory of judgment.

Frege therefore begins his analysis of general propositions by rejecting the traditional subject–predicate account of judgments. He substitutes for it an account in terms of 'function' and 'argument,' and this account in turn yields an analysis of general sentences.

The need for such a radical revision had dawned on him slowly. 'In the first draft of my formula language I let myself be misled through the example of language to compose judgments out of subject and predicate,' he writes. 'But I soon convinced myself that this was an obstacle for my particular purpose and that it led only to unnecessary complications' (BS, p. 4). If logicians had accepted the assumption that judgments were composed out of subject and predicate, they had simply been misguided by the structure of the Indo-European languages. Frege explains that in his logic 'the departures from the tradition have their explanation in the fact that logic has so far tied itself too closely to language and grammar' (*ibid.*, p. xiii). The purpose of the *Begriffsschrift* was to design 'a formula language of pure thought' and for that purpose the grammar of historically grown languages could be no guideline.

A LANGUAGE OF PURE THOUGHT

In the attempt to construct such a language there arises a distinction which remains important for the whole of Frege's later development. He thinks that in the utterances of ordinary language we must distinguish the psychological intentions and associations of speaker and hearer from the actual objective content of the utterance itself. Later on Frege speaks of the 'illumination' surrounding a word or sentence, the 'coloring' it possesses for speaker or hearer. These are merely subjective, associative, psychological features which have no place in a language of pure thought.

> All phenomena of language which arise out of the interaction of speaker and hearer (in that, e.g., the speaker takes into account the expectations of the hearer and tries to put them on the right track even before uttering the sentences) have no equivalent in my formula language (*ibid.*, p. 3).

He tries to show that subject and predicate are distinguished on the basis of subjective attitudes and expectations of speaker and hearer; they are not objective logical features. What gets identified as the subject of a judgment is that which the speaker singles out as deserving particular attention. 'One can only say: "Subject is that concept of which the judgment speaks predominantly." The place of the subject in the series of words has the meaning of a *designated* position in language, to which one wants to guide the attention of the hearer' (*ibid.*).

The objective conceptual content of the judgment, which Frege's logic strives to symbolize, is that 'which influences possible logical consequences' and no more. 'Everything that is necessary for a correct chain of inferences will be completely expressed . . . nothing is left to guessing' (*ibid.*). Logic for Frege is concerned with the study of the logical relations between the objective contents of judgments. That these judgments occur in concrete speech situations and are there surrounded by a host of associated phenomena is of no relevance. On the basis of this separation the subject–predicate distinction is supposed to be shown up as irrelevant to logic.

> In order to justify this I remark that the contents of two judgments may differ in two ways: either the consequences derivable from the first when it is combined with certain other judgments always follow from the second when it is combined with the same judgments, or this is not the case. The two propositions 'The Greeks defeated the Persians at Plataea' and 'The Persians were defeated by the Greeks at Plataea' differ in the first way. . . . Now I call that part that is the same in both the *conceptual content*. Since it alone is of significance for our *Begriffsschrift* we need not introduce any distinction between propositions having the same conceptual content (*ibid.*, pp. 2–3).

A LANGUAGE OF PURE THOUGHT

5 Frege's logical notion of a function

In place of the subject–predicate distinction Frege introduces that of function and argument. 'In this I follow the example of the mathematical formula language where one would do violence if one tried to distinguish subject and predicate' (BS, p. 3). The notion of function adopted here has its origins in mathematical analysis, but Frege generalizes it for his own logical purposes (*ibid.*, p. 19). He hopes that his new account of judgment will prove to be of lasting value because, as he says, 'one can easily recognize how the view of a content as a function of an argument leads to the formation of new concepts' (*ibid.*, p. xiii). It does so, presumably, by providing us with a more flexible tool for the analysis of propositions. The traditional account understands every sentence as ascribing something to a subject. Frege's new account treats propositions that ascribe properties to individual objects as special cases of ones in which an n-place function is applied to n arguments. A subject–predicate proposition contains a one-place function with one argument; a proposition expressing a relation between two objects contains a two-place function and two arguments. The logic of predication and the logic of relation are thus united under one heading. There are other ways in which the function–argument distinction makes our analysis of propositions more flexible. The same proposition $\phi(a,b)$ can at one time be regarded as a function of an argument a, at another time as a function of an argument b, and at yet another time as a function of both a and b. And again, if $\phi(a)$ is generally regarded as a function of a, we can also at other times consider it a function with ϕ as argument (*ibid.*, pp. 18–19). And the most important way in which the function–argument account gives rise to new concepts is as the basis for an account of general propositions.

In the mathematical usage of Frege's time the terms 'function' and 'argument' refer to certain kinds of *expression* within a mathematical formula. According to Euler's classical definition, 'a function of a variable quantity is an analytic expression which is composed in a certain way out of the variable quantity and constant numbers or quantities.'[21] This characterization shows both that a function is taken to be an expression rather than what an expression stands for, and also that this conception is hardly carried through consistently since the function is also said to be composed of numbers or quantities. There is, then, in the mathematical usage of Frege's time a peculiar wavering in the use of the term between what might be called its syntactic and its semantic sense.

In the *Begriffsschrift* Frege completely follows that imprecise mathematical usage. When he introduces the distinction of function and argument he makes it clear that it is meant as syntactical. Functions and arguments are for him parts of sentences. The distinction 'has nothing to do with the conceptual content; it comes about only because we view the expression in a certain way' (BS, p. 15). The formal characterization is given purely syntactically:

> If in an expression . . . a simple or complex sign occurs in one or more places, and we consider it replaceable in one or more of these places by something, but everywhere by the same thing, then we call the part of the expression which hereby appears as unchangeable the function and the replaceable part its argument (*ibid.*, p. 16).

This usage differs from Frege's later use of the terms 'function' and 'argument,' for later on these are taken to be the references of certain kinds of expressions. What he earlier called 'functions' then come to be called 'functional expressions' and what he called 'arguments' become 'argument signs.'

Frege's use of the distinction goes beyond the mathematical one in having a radically wider range of application. Functions are for him not just calculating expressions, they also can serve to express numerical identity (as he says in the essay 'Applications of the *Begriffsschrift*' of 1879, BS, p. 90), concepts (e.g., 'is representable as the sum of four squares' *ibid.*, p. 17), and relations (e.g., 'killed,' *ibid.*, p. 16). It has sometimes been maintained by interpreters (including Dummett) that the identification of concepts and relations with functions is a later addition to Frege's theory and derives from the semantic doctrine (developed around 1891) that declarative sentences are names of truth-values. But this is surely mistaken and completely underestimates the central logical role of the notion of function in Frege's thought from the pre-*Begriffsschrift* days onwards. What changes around 1891 is only the explanation of why concepts and relations are functions, but not the identification itself.[22]

The distinction of function and argument in an expression which Frege initially considers as merely syntactical can gain semantic significance under certain circumstances. Because of this Frege also allows himself to speak occasionally of seeing 'the *content* as a function of an argument' (*ibid.*, p. xiii).

The distinction to begin with is one between the part of an expression that is considered fixed and the part that is considered variable. He adds that, as long as function and argument are *in fact* determinate in a particular sentence, the distinction remains purely syntactical and has no logical importance.

> But when the argument becomes *indeterminate* as in the judgment: 'You can take as argument for "being representable as a sum of four squares" any arbitrary positive whole number: the sentence will always be correct,' then the distinction of function and argument takes on significance for the content. . . . Through the distinction of the

determinate and the *indeterminate*, . . . the whole is divided into function and argument not only because we view it that way, but also in its content (*ibid.*, p. 17).

In a complex expression we can replace an argument by a letter (a 'variable,' in a terminology which Frege dislikes) and thereby indicate the indeterminateness. If the original expression was a sentence we now have a sentence form that can be transformed into a series of other sentences through suitable substitutions for the letter.

6 *The analysis of general sentences*

According to Frege's account, sentence forms such as those just described occur in general sentences. The distinction of function and argument is therefore logically significant for such sentences. To say that 'Everything is a ϕ' is to say that 'the function is a fact whatever one might take as its argument' (BS, p. 19). A general statement is thus a statement about a function. From the way Frege speaks in the *Begriffsschrift* it often appears that a general statement, like a statement of sameness of content, is a statement about the signs rather than about what the signs stand for. If a general statement is a statement about a function, then according to the definition it is a statement about an expression. But the looseness of the terminology also allows us to consider a general statement as a statement about the content expressed by the functional sign. This indeterminateness is characteristic of Frege's early writings and is cleared up only after 1884 when a general statement is always taken as a statement about a function in the second sense.

In any case, according to the *Begriffsschrift* account the statement 'All men are mortal' is not a statement about all men, but one about the function 'if a is a man, then a is mortal,' saying of it that it is a fact whatever we substitute for 'a.' The sentence 'Some men are Greeks' does not make a statement about some men, but does about the function 'a is a man and a is Greek,' saying of it that not all sentences resulting from substitutions for the letter 'a' are false. 'Everything is mortal' says that the expression 'a is mortal' turns into a true sentence whatever we substitute for 'a' and 'There are men' says that the expression 'a is a man' turns into a true sentence for some substitution for the letter 'a.'

The claimed generality may apply to a whole sentence form or only to part of it. An appropriate notation, Frege argues, must indicate the exact scope of a sign for generality, that is, it must show to which part of a sentence the generality claim is supposed to apply. Our notation must also be flexible enough to allow substitution of letters for more than one argument in a sentence, and thus the forming of multiply general sentences.

A LANGUAGE OF PURE THOUGHT

The Fregean analysis of general sentences therefore involves the combination of three elements:

(1) the distinction of function and argument which permits him to distinguish between a statement about a particular argument and a statement about a function;
(2) the use of letters (or variables) to express the distinction between determinate and indeterminate places in an expression, a device taken from arithmetic;
(3) the introduction of a notation that allows precise indication of the scope of a claim of generality and the distinction of different argument places in one sentence with different attached claims of generality.

It is Frege's view that ordinary language suffers in each of these respects. It treats general sentences as if they made statements about objects (e.g., all men or some men), rather than functions; it lacks an adequate notation for indeterminate places; and it cannot express very precisely the scope of a generality claim. Ordinary language is adequate for speaking more or less determinately of the things we need to communicate about in everyday life, but it is insufficient for the precise expression of logical and mathematical principles.

In a letter to Peano written after 1891 Frege points out that 'among most logicians there exists a great lack of clarity about the character of existential judgments, as much among the followers of Boole as among the psychological logicians, in spite of the fact that Kant seems already to have been on the right track in his critique of the ontological argument for the existence of God' (WB, p. 176).[23] Kant and Frege agree in the *negative* thesis that existence is not a predicate or property of a thing. In the *Critique of Pure Reason* Kant wrote: '"Being" is obviously not a real predicate; that is, it is not a concept of something which could be added to the concept of a thing. It is merely a positing of a thing, or of certain determinations, as existing in themselves' (A 598). The agreement between Kant and Frege reveals itself in the fact that both of them conclude that the ontological argument for the existence of God cannot be valid, since existence cannot be one of the perfections of God.[24]

But there is clearly no complete agreement between Kant and Frege. Kant says that in an existential judgment we 'posit' a thing, whereas in Frege's account there is no recourse to this obscure notion. Instead, an existential judgment is taken to be a statement about a function. The difference between the two conceptions is clearly brought out by their different consequences. For Kant, *no* existential judgment can be analytic. That indeed seems to have been one of the reasons for his belief that arithmetical truths are synthetic *a priori*; only intuition can furnish us with the objects necessary to make the existential claims of arithmetic true. But it is precisely one of Frege's goals to show the analyticity of arithmetic. He must therefore hold that existential judgments *can* be analytic.

Frege does not say, as Kant does, that 'being' is no real predicate, but only that it is not a predicate of an object. For him it is rather a predicate of a function, a second-level predicate in his later terminology. Kant's doctrine allows as well formed both the propositions 'Horses exist' and 'This table exists,' but claims that in neither is a real predicate being predicated. In the one proposition horses are posited, in the other this table. It is a consequence of Frege's account, on the other hand, that 'This table exists' is illegitimate, since in it the expression for a particular object is combined with the term 'exists,' which stands for a second-level function that requires a first-level function as argument. Nevertheless, it *seems* as if such formulations are meaningful. It seems, for instance, perfectly meaningful to say of somebody whose name is Sachse that 'Sachse exists.' To this Frege replies: 'If "Sachse exists" is supposed to mean "The word 'Sachse' is not an empty sound, but signifies something",' then there is no objection to the formulation. It expresses only a self-evident presupposition for all our words (NS, p. 67). 'But if the sentence "Leo Sachse is" is trivial, then there cannot be the same content in the "is" as in the "there are" of the sentence "There are human beings," for the latter does *not* say something trivial' (*ibid.*, p. 69). If one is tempted to translate the sentence 'There are human beings' into 'Some human beings exist,' the basic fallacy of such reasoning is to think that the real predicate of the sentence is contained in the word 'exist.' In the sentence in question that word 'is a mere form word which must be understood similarly to the "it" in "it is raining." In the same way as language invented the "it" because of its need for a grammatical *subject*, so it also invented the "exist" because of its need for a grammatical *predicate*' (*ibid.*).

> The copula is excellently suited for the formation of a concept without content, i.e., for the mere form of the predicate without content. In the sentence 'The sky is blue' the predicate is 'is blue,' but the real content of the predicate lies in the word 'blue.' When one omits that word a statement without content is left: 'The sky is.' Thus, one forms the quasi-concept 'being' which is without content because it has an unlimited extension. Now one can say: Human beings = existing human beings. 'There are human beings' is the same as 'Some human beings are' or 'Some being is human.' Here the real content of the assertion does not lie in the word 'being,' but in the form of the existential judgment. . . . When philosophers speak of 'absolute being' they are really only deifying the copula (*ibid.*, p. 71).

The Fregean analysis of general (i.e., universal and existential) judgments is supposed to permit the precise formal analysis of arithmetical notions and laws, but its implications are more general. It is also supposed to destroy certain misconceptions to which the notions of being and existence give rise in philosophy.

There remains the question whether the notion of existence as a second-level concept fully captures the traditional notion of existence. The Fregean notion, it must be remembered, is comparatively thin in content. To say that something exists is in Frege's account no more nor less than to say that a concept has instances or that the concept can be predicated truly of an object. Frege therefore distinguishes existence in his sense from actuality. While existence is for him a second-level concept, actuality is taken by him as a first-level predicate, a predicate of objects. To say that something is actual is to say that it exists in a spatio-temporal field and that it enters into causal chains. Some of the problems philosophical ontology has dealt with in the past are concerned with this notion of actuality. For instance, when we ask what individuates a real spatio-temporal object or under what conditions we would identify different temporal stages as belonging to the same object, the questions that arise concern our notion of actuality, not the Fregean notion of existence. For Frege what is actual or real is to be contrasted with what is objective but not actual or real. The things logic deals with characteristically are objectively unreal. The task of logic is the investigation not of the notion of actuality, but of existence and objectivity. It follows, then, that not everything philosophers have discussed under the heading of ontology is part of the subject matter with which Frege is concerned.

7 *The priority of judgments over concepts*

Substituting the distinction between function and argument for that between subject and predicate means more than the adoption of a flexible terminology for Frege. With it goes a view of the relation of the judgment to its parts that is completely different from that of traditional logic.

The viewpoint of subject–predicate logic was essentially aggregative. It saw the judgment as formed by aggregation out of previously given constituent concepts. This conception led to the confusion between those combinations of terms which form judgments and those which form only complex concepts. Traditional logic (including that of Leibniz) was given to identifying judgments and complex concepts. Only Kant was able to cut the Gordian knot (while maintaining the terminology of subject and predicate) by seeing that judgments possess a peculiar kind of unity. Kant argued that there could not be any combination of ideas unless there were already an original unity that permitted such combination. The variety and complexity of our judgments presuppose a transcendental unity of consciousness which immediately gives rise to and is reflected in the unity of the judgments themselves. Our use of concepts presupposes this original unity of judgment. 'The only use which the understanding can make of these concepts is to judge by means of them' (*Critique*, A 68). Kant spells this out in words which anticipate the Fregean doctrine of concepts:

> Concepts, as predicates of possible judgments, relate to some representation of a not yet determined object. Thus the concept of body means something, for instance, metal, which can be known by means of the concept. It is therefore a concept solely in virtue of its comprehending other representations, by means of which it can relate to objects. It is therefore the predicate of a possible judgment, for instance, 'every metal is a body' (*ibid.*, A 69).

Concepts for Kant are essentially predicative; they can be understood only through their role in judgments. Judgments possess a unity and thereby a logical form that is independent of their content. Concepts therefore cannot be the mere product of abstraction from sensory experience. They must already contain an *a priori* element.

The doctrine of the priority of judgments over concepts has two distinct functions in Kant's philosophy. It is meant to show the impossibility of atomism and to undermine empiricism. Kant was familiar with atomism in two quite different forms: Leibniz's atomism, which claimed to analyse concepts into simples and substances into monads, and Humean atomism, which held that knowledge is composed of atomic sensations. While both rationalist and empiricist versions of atomism are supposed to be refuted by the doctrine of the priority of judgments, it counts more strongly against empiricism. Kant's forms of judgment are determined *a priori* by logic. The priority of judgments over concepts thus implies that every empirical concept presupposes pure concepts of the understanding.

In his reply to Schröder's attack Frege maintains that his logic differs from Boole's in basic design. 'This is due to the fact that I have moved further away from Aristotelian logic. For in Aristotle just as in Boole the formation of concepts through abstraction is the fundamental logical operation and judging and inferring are brought about through direct or indirect comparison of the extensions of these concepts' (NS, p. 16). Frege does not mean to say that Aristotle and Boole are mainly concerned with concept formation; on the contrary, their logics are predominantly theories of inference. But their theories are based on a certain view of concept formation which remains unexamined and is in fact unsatisfactory. The tradition to which Aristotle and Boole belong mistakenly assumes that concepts are formed by abstraction from individual things and that judgments express comparisons of concepts, as do inferences. This tradition treats concepts as if they were initially independent of judgments and entered them only incidentally. 'In contrast to Boole,' Frege wrote, 'I begin with judgments and their contents and not with concepts.

... The formation of concepts I let proceed from judgments' (*ibid.*, p. 17).

The view which Frege expounds here implies at least four different things: (1) an epistemic claim concerning the contents of judgments, (2) an account of the relation of judgments to concepts, (3) a thesis about the nature of concepts, and (4) a methodological principle for meaning analysis.

(1) The first implication is that the contents of judgments are epistemically primary. In 1882 Frege wrote to a correspondent:

> I do not think that the formation of concepts can precede judgment, for that would presuppose the independent existence of concepts; but I imagine the concept originating in the analysis of a judgeable content. I do not think that for every judgeable content there is only one possible analysis or that among the various possibilities one can always claim objective precedence (WB, p. 164).

The claim that judgments are epistemically and logically prior to concepts is reformulated by Frege in the summary of his major doctrines which he wrote for Darmstaedter in 1919:

> My particular conception of logic is initially characterized by the fact that I put the content of the word 'true' at the beginning and let it be followed immediately by the thought with respect to which the question of truth arises. In other words, I do not begin with concepts out of which the thought or judgment is composed, but I get to the parts of the thought through its analysis. In this respect my *Begriffsschrift* differs from similar creations by Leibniz and his successors, in spite of its possible unhappy name (NS, p. 273).

The term '*Begriffsschrift*' is unhappy if it is taken to mean that concepts rather than judgments are primary. That was of course not the sense in which Frege had originally used the term. The term '*Begriffsschrift*' was meant to imply that logic dealt with objective conceptual contents and not with subjective psychological facts of thinking. It was not meant to deny the primacy of the judgment. In that sense Frege writes against Schröder: 'In this connection it is remarkable that some recent linguists consider the sentence-word the basic form of speech' (*ibid.*, p. 19).

(2) The doctrine implies a certain view of concepts. Simple concepts and relations 'originate together with the first judgments in which they are ascribed to things' (*ibid.*). Concepts are always reached through the splitting up of judgments, through analysis; they are not given separately and the judgment is not composed out of previously given constituents. As elsewhere, Frege here rejects the view that thinking is an aggregative process. 'I doubt whether there exists any thought whatsoever answering to this description' (F, p. iii).

If we consider concepts as something ready-made out of which judgments are composed, we end up with a logic like Aristotle's or Boole's. For them the formation of new concepts is always a question of new (conjunctive or disjunctive) combinations of previously given concepts. 'With this form of concept-formation one must presuppose as given a system of concepts or, metaphorically speaking, a network of lines. In this the new concepts are really already contained' (*ibid.*, p. 38).

The aggregative viewpoint makes the formation of concepts a mechanical process. But Boolean logic 'represents only part of our thinking; the whole can never be carried out by a machine or be replaced by a purely mechanical activity' (*ibid.*, p. 39). Aristotle and Boole assume perfect concepts and therefore take the most difficult task for granted. One can indeed then draw conclusions out of premises by means of a mechanical calculating procedure, but a complete logic requires more. There must be a method of concept-formation that can produce scientifically fruitful concepts with completely new boundary lines (*ibid.*; cf. also Γ, p. 100).

From these words against Schröder it is clear that Frege assumes he has shown in the *Begriffsschrift* how such a task is to be accomplished. Through analysis of a sentence he has reached the notions of function and argument and through them in turn he has discovered a logically more adequate theory of judgment and a satisfactory account of general propositions. Logic is for Frege not just a question of mechanical (recursive) operations on concepts and symbols. It also involves analysis, a non-mechanical, cognitive process. In his *Logik* (1880, p. 479) Lotze had explained this basically Kantian view of the interdependence of analytic and synthetic methods:

> The analytic method is essentially a process of investigation which seeks to discover truth, the synthetic a process of presentation which seeks to transmit, in its own objective context, a truth that has somehow been found.

(3) A further point implied by Frege's doctrine is also straightforwardly derived from Kant. He concludes from the relation of judgment and concept that a concept 'is nothing complete, but only a predicate of a judgment, for which a subject is still lacking' (NS, pp. 18–19). The sign for a property therefore never occurs in the *Begriffsschrift* 'without at least indicating a thing which might have the property, the sign for a relation never without indicating things that might stand in that relation' (*ibid.*, p. 19). The doctrine of Frege's later essay 'On Function and Concept' is present here in a nutshell: functions, concepts, and relations are incomplete and require variables in their expression to indicate places of arguments. The doctrine is here brought together with that of the priority of judgments over concepts.

(4) The final point is that in a logically perfect language we require the

signs making up the expression of a judgment to correspond exactly to the parts of the judgment which logical analysis has distinguished. In a language that is not logically perfect we cannot safely make that assumption. In ordinary language we can presuppose only that the sentence as a whole expresses a thought. Whether or not the parts of the sentence have a meaning is open to question; and the meaning we assign to such parts will depend on the meaning of the whole sentence.

> The task of natural language is essentially fulfilled when the people that communicate in it connect the same thought with the same sentence, or at least approximately the same. For that it is completely unnecessary that the words taken by themselves have a sense and a reference, as long as the sentence as a whole has a sense. Things are different when inferences are supposed to be drawn. Then it is essential that the same expression occur in two sentences and that it have exactly the same reference in both. The expression must therefore have a reference that is independent of that of the other parts of the sentence (KS, p. 236).

With respect to some words of ordinary language we can safely assume that they signify something on their own, but with respect to others we can only assume that they mean something in the context of the sentence. This idea already guides Frege's account of general propositions in the *Begriffsschrift*. Comparing the two sentences

(a) The number 20 can be represented as the sum of four squares,

and

(b) Every positive integer can be represented as the sum of four squares,

Frege argues that it is a mistake to consider the expressions 'the number 20' and 'every positive integer' as the subjects of their respective sentences, for they are

> not concepts of the same rank. What is asserted of the number 20 cannot be asserted in the same sense of 'every positive integer.' . . . The expression 'every positive integer,' unlike the expression 'the number 20,' does not yield an independent idea but acquires a meaning only from the context of the sentence (BS, p. 17).

In the *Foundations of Arithmetic* Frege states as an explicit methodological principle 'never to look for the meaning of a word in isolation, but only in the context of a sentence' (F, p. x). In this form the doctrine has become known as Frege's context principle. It is meant here primarily as a methodological principle for the analysis of sentences of ordinary language. But the implication seems to be that there is a priority of sentence meaning over word meaning for every language, including a logically perfect one. In this stronger sense the principle would amount to the reaffirmation of the Kantian doctrine of the priority of judgments over concepts.

The context principle has been the source of much confusion and disagreement in the interpretation of Frege's writings. Some, like Carnap, have ignored it altogether; others have tried to give it a central place in their reading of the text.[25] Michael Dummett has tried to show that the principle must be interpreted weakly, rejecting its supposed epistemic implications. He has also claimed that Frege abandoned it after he came to develop his sense-reference semantics. The evidence seems to contradict both of Dummett's claims. His attempt to associate Frege with contemporary (recursive) model-theoretic semantics appears to be his only reason for the first, and an inadequate consideration of the historical setting the reason for the second.

The doctrine of the priority of judgments over concepts can be understood only if it is seen as deriving from deep features of Frege's thought. It expresses one of the Kantian elements in his thinking. Together with the Leibnizian idea of a perfect language and that of the reduction of arithmetic to logic these elements constitute the guiding principles for the construction of the *Begriffsschrift*. The *Begriffsschrift* logic is constructed not just as a pragmatic-technical device. It expresses a definite philosophical viewpoint, as Frege makes clear in his response to Schröder. We might say that it is only in the context of this philosophical outlook that the context principle has its meaning. In that context its meaning is not exhausted by the weak interpretation which Dummett has given of it. And assuming it is indeed anchored deeply in Frege's thought, it is implausible to conclude with Dummett that in his later years Frege simply let it slip from his mind.

When we step back from the particular issues that have so far concerned us and ask why Frege was interested in language and the construction of a logical language we discover that his interest was never an end in itself. He was, to use Ian Hacking's apt distinction, not interested in the *pure* theory of meaning, but in *applied* philosophy of language.[26] Thinking about language was essential for any real improvement in logic. And Frege considered such an improvement both possible and desirable. But even the interest in logic was not philosophically ultimate for him, we discover on closer inspection. Even that interest was in Hacking's sense not pure, but applied.

IV

In Search of Logical Objects

1 Frege's methodological critique of mathematics

Frege conceived the new logic not as an end in itself but as a tool for the analysis of arithmetic. 'I intend to apply it first of all to that science,' he wrote in the preface of the *Begriffsschrift*, 'attempting to provide a more detailed analysis of its concepts and deeper foundations for its theorems' (BS, p. xiv). This purpose, he felt, separated his logic from Boolean algebra which remained 'within the boundaries of pure logic' (NS, p. 15) and was therefore incapable of expressing actual contents (*ibid.*, p. 39).

While working on the *Begriffsschrift* he was fully convinced that the new logic would eventually provide the desired analysis of arithmetic. 'To proceed further along the path indicated,' he wrote, 'will be the object of . . . investigations which I shall publish immediately after this booklet' (BS, p. xiv). In fact it took him another five years until he succeeded in pushing the project any further in the *Foundations of Arithmetic*. And what he then published was not exactly a continuation of the *Begriffsschrift*. The earlier work had concluded with the definition of the notion of a series, but the *Foundations* did not simply continue from that point. It did not even make use of the symbolism of the *Begriffsschrift*. Instead, it examined in everyday language various philosophical views about numbers and arithmetic and, in the second half of the book, proposed a more adequate definition of the natural numbers and the concept of natural number.

The reasons for the departure from the original plan are not difficult to understand. As Frege began to think about the logical analysis of arithmetic he came to realize more and more the lack of agreement among mathematicians about the meaning of the most elementary terms of their science. Recapitulating his views at the end of his life, he wrote:

> With almost every technical term of arithmetic . . . the same questions recur: Is the visible that with which arithmetic is concerned or is the

visible only a sign for the subject-matter of arithmetic, an auxiliary device and not the object of investigation? . . . Until arithmeticians agree in their answers to these questions and until their terminology consistently corresponds with their answer, there is really no arithmetical science, unless that science is a totality of verbal sounds quite apart from the sense of those sounds or from even their having any sense at all (NS, p. 277).

In particular it is the concept of number about which mathematicians are in utter confusion, he argued. 'Perhaps arithmeticians sometimes only pretend to understand by "number" what they claim to understand. If that is not the case, then they connect different meanings with sentences that sound the same. And if they still believe themselves to be working on the same science, then that is merely an illusion' (*ibid.*, p. 276).

The methodological failures of mathematics are a recurring theme in Frege's writings. They are touched upon repeatedly in the first half of the *Foundations of Arithmetic*; they are taken up again in an essay entitled 'Logical Flaws in Mathematics,' written around 1899 (NS, pp. 170–81); they preoccupy him throughout most of the second volume of the *Grundgesetze;* and they are pursued most systematically in an essay on 'Logic in Mathematics' of 1914 (*ibid.*, pp. 218–70). Mathematicians, he repeats over and over again, are confused about the meaning of simple terms such as 'number,' 'set,' 'equality,' 'variable,' and 'function.' And their confusion is systematic; it reveals a coherent pattern. In the *Foundations of Arithmetic* (pp. ii–iii) Frege writes:

> If a concept fundamental to a mighty science gives rise to difficulties, then it is surely an imperative task to investigate it more closely until those difficulties are overcome. . . . Many people will certainly think this not worth the trouble. They naturally suppose that this concept is dealt with adequately in elementary textbooks where the issue is settled once and for all. Who can believe that there is anything still to learn on such a simple matter. . . . The first prerequisite for learning anything is thus utterly lacking — I mean the knowledge of our ignorance.

The most naive and unreflective answers as to what a number is are taken for granted without any thought being given to the methodological problem of how the meaning of an expression is to be determined. A number term is taken out of the context in which it occurs and the question is asked what the term means. The answer that emerges from this procedure is that numbers are internal psychological objects or are simply the number signs themselves. Psychologism and formalism in mathematics are both the offspring of this defective methodology.

Mathematicians say that numbers are abstracted from classes or sets; but they fail to give a coherent account both of what they mean by abstraction and of what they mean by 'class' or 'set.' If the notion of abstraction means anything it means that we abstract when we compare objects and discover that they agree in some properties but not in others.[1] When we abstract we proceed from objects to a concept under which those objects fall. But mathematical abstraction is supposed to be something quite different. Cantor, for instance, wished to create the numbers by abstraction. According to him we create the ordinal number five by abstracting from the particular properties of a set of five points on a line, and, by abstracting once more from the order of the elements, we obtain the cardinal number five. Frege thinks Cantor has a completely magical view of abstraction; for abstraction can generate only concepts, not objects like the numbers (NS, pp. 77–8).

There is, according to Frege, an easy criterion for showing that the usual characterizations of numbers are barren. If a mathematician defines a number as an idea, a sign, a series of similar objects, or as something reached by abstraction from sets of particulars, we merely need to ask whether such 'definitions' have any use in the construction of mathematics as a system. If they cannot be applied literally or if their literal application does not lead to the ordinary laws of arithmetic, they should be thrown away as useless. Such definitions do not specify the meanings number terms possess in the context of the theory.

> When we peruse mathematical writings we come across many things that look like definitions and are also called definitions without being such. Such definitions can be compared to the kind of stucco decorations on buildings that look as if they carried weight but could be removed without affecting the structural safety of the building. Such definitions can be recognized by the fact that they are not needed, that they are never called upon in a proof (NS, p. 229).

Philosophical confusions of this kind generate many others, such as confusions about the meaning of the identity sign. 'The view of number as a series of similar objects, as a herd, as a heap, or as a whole consisting of similar parts is closely connected with the view that the equality sign is not used to express identity' (*ibid.*, p. 244). For it might then be that each occurrence of '5' in '5 = 5' refers to a different series of objects and, hence, '=' cannot mean identity.

> Mr. G. Peano says that the opinions of writers about the identity sign differ remarkably, and unfortunately he is right. That is to say something most significant. If we remove identity from arithmetic, there is barely anything left of it. Peano's comment therefore means that mathematicians are of different opinions concerning the meaning of

most of their theorems. A non-mathematician would be appalled to hear that this is not considered particularly bad, that mathematicians believe they have more important things to do than to try to remove this hair-raising state of affairs (*ibid.*, p. 180).

This type of confusion is also evident in the usual mathematical talk about variables and functions. Mathematicians speak as if variables referred to variable or indefinite numbers in the same way in which particular numerals refer to constant and definite numbers. The name 'variable' itself is misleading here, and Frege advocates its elimination from mathematics. For him the reference of an expression is always something constant and definite. He considers talk about variable and indefinite numbers to be a result of the naive belief that every expression in the mathematical symbolism functions in the same way. More careful attention to the way in which 'variables' are used, however, reveals that they function quite differently from particular number terms. They function in two ways. In some expressions they indicate an open place into which a constant can be substituted, as in '$x - 2$.' In other contexts they indicate generality, as in '$x + y = y + x$.' In both cases, Frege says, the function of the variable is to 'indicate'; in neither case has it got a reference (cf. NS, p. 269).

Mathematical confusion about variables carries over into confusion about the notion of a function, since functional expressions usually contain variables. Frege quotes a characteristic mathematical description of functions: 'If every value of the real variable x that belongs to its range has correlated with it a definite number y, then in general y is also defined as a variable and is called *a function of the real variable x*. This relation is expressed by an equation of the form $y = f(x)$' (KS, p. 276). From such descriptions it appears that functions, like variables, refer to variable or indefinite numbers or magnitudes. But there are no such numbers, and functions must be distinguished sharply from the numbers. 'Mathematics should really be a model of logical clarity. In fact there are perhaps more wrong expressions in mathematical writings than in those of any other science and as a result more wrong ideas' (*ibid.*, p. 280).

There are several methodological failures in the usual mathematics. First, there is the tendency of mathematicians to confuse the sign with what it signifies. 'We discover here a widespread mathematicians' disease which I would like to call *"morbus mathematicorum recens."* Its main symptom consists in the inability to distinguish between a sign and the signified' (NS, p. 241). In the *Begriffsschrift* Frege himself had treated that distinction lightly, but after 1884 he always insists that it should be made completely clear in all contexts. For that purpose he suggests a systematic use of quotation marks in order to distinguish a sign for an object from a name for the sign itself.

Mathematical discourse about the most elementary concepts is also confused because it fails to draw two other fundamental distinctions between concept and object, and between the subjective or psychological and the objective or logical (cf. F, p. x). Finally, when mathematicians examine the meaning of their terms they tend to consider them in isolation, but 'one must ask for the meaning of words in the propositional context, not when they are isolated' (*ibid.*).

The tendency of mathematicians to isolate their elementary terms when they examine their meaning is an expression of a deeper failure. They do not see that mathematics really constitutes a system. It is not sufficient in mathematics to prove as many theorems as possible or to state as many true propositions as one can find. It is more important to show how theorems and concepts are interrelated. In the essay 'Logic in Mathematics' Frege writes:

> Only in a system is a science complete. One can never abandon this need for a system. Only through a system can one gain complete clarity and order. No science can control its material as well as mathematics can and work it through until it becomes completely transparent; but no other science can lose itself in such a heavy fog as mathematics can when it abandons the construction of a system (NS, p. 261).

The meaning of a mathematical term is not the complex of ideas and associations that may be connected with it in our minds. Its meaning is its role in the mathematical theory. Similarly, the meanings of mathematical propositions do not consist in the thoughts they might conjure up in the mind, but in their place in the mathematical system. Euclid understood this point when he set out to axiomatize geometry. The purpose of logical analysis is to continue the work started by Euclid. Only by making explicit what is usually taken for granted is it possible to trace chains of inference from basic truths to more and more complex theorems. Only then is it possible to define the basic terms in a functional and therefore satisfactory manner.

> When one asks wherein lies the value of mathematical knowledge the answer must be: it lies less in what is known than in how it is known, less in the subject matter than in the extent of its illumination and in the insight it provides into the logical network (*ibid.*, p. 171).

2 *The strategy of Frege's anti-empiricism*

Frege's conviction that mathematics is methodologically deficient in the treatment of its own foundations determines the structure of the *Foundations of Arithmetic*. The task of the book is to settle the question whether the natural numbers and the concept of the number are definable (F, p. 5). That question arises, in the first instance, out of the internal development of mathematics itself where 'proof is now demanded of many things that formerly passed as self-evident' and definitions are given of terms that were formerly understood intuitively (*ibid.*, p. 1). 'In all directions these same ideals can be seen at work—rigor of proof, precise delimitation of extent of validity and, as a means to this, sharp definition of concepts' (*ibid.*).

In order to determine the meaning of number terms it is first necessary to clear away certain misconceptions about numbers. Only then can one proceed to the definition of number and numbers. The course of the *Foundations of Arithmetic* is therefore set by just this requirement.

Rigor and precision are not Frege's only goals; the clearing up of methodological confusions is not his only concern. 'Philosophical motives too have prompted me to inquiries of this kind. The answer to . . . questions . . . about the nature of mathematical truths—are they *a priori* or *a posteriori*, synthetic or analytic?—must lie in this same direction' (*ibid.*, p. 3). If the numbers are not definable it is unlikely that the laws of arithmetic could be analytic truths. The question of the definition of the numbers is therefore of philosophical as well as technical and methodological interest. 'On the outcome of this task will depend the decision as to the nature of the laws of arithmetic' (*ibid.*, p. 5).

The philosophical purpose of the *Foundations of Arithmetic* is to establish beyond any possible doubt that arithmetical truths are *a priori*. If this fact can be established, Kant's rejection of empiricism will be vindicated. As Frege says (in a passage previously quoted): the reception of his results 'will probably be worst among those empirics who want to recognize only induction as an original form of inference and even that not as a form of inference but as habituation. Perhaps one or the other of them will test once more the principles of his epistemology on this occasion' (*ibid.*, pp. x–xi).

The claim that arithmetical truths are *a priori* is plausible because the basis of arithmetic is said to lie deeper than any empirical truth. Our empirical laws are discovered by means of induction, but induction itself is dependent on arithmetic. 'The procedure of induction can presumably only be justified by means of general propositions of arithmetic. . . . Induction must lean on the theory of probability. . . . But how that theory could possibly be developed without presupposing arithmetical laws is beyond comprehension' (*ibid.*, pp. 16–17).

In the *Foundations of Arithmetic* Frege's aim is to show not only that arithmetical truths are *a priori*, but also that they are analytic, logical truths. 'The arithmetical truths govern the domain of the countable. That is the widest of all: not only what is actual or can be intuited belongs to it, but everything thinkable' (*ibid.*, p. 21). Because arithmetic deals with everything thinkable, it is really logic. He concludes his book by saying: 'From all the preceding it thus emerged as a very probable conclusion that the truths of arithmetic are analytic and *a priori*; and we achieved an improvement on the views of Kant' (*ibid.*, pp. 118–19). It is an improvement on the view of Kant rather than a refutation of his doctrine that arithmetical laws are synthetic *a priori,* because it supports Kant's claim that arithmetic is *a priori* and that is his fundamental insight. Kant was 'a great mind we can only look up to with grateful admiration' (*ibid.*, p. 101). He set out to show that not all knowledge is empirical, that human thinking is not just a subjective, psychological process, that it consists in the grasping of objective thoughts, and that empiricism and naturalism are untenable.

Anti-empiricism is in fact pervasive in Frege's book. It is the guiding philosophical thought in his critique of other conceptions of the foundations of mathematics. Frege attacks the thesis that arithmetical laws are inductive (*ibid.*, p. 12), that numbers are properties of external things (*ibid.*, p. 27), and that they are subjective mental entities (*ibid.*, p. 33). And he also argues against the assertion that they are 'agglomerations' of physical objects (*ibid.*, p. 38). The formalist claim that numbers are to be identified with number signs is briefly dismissed (*ibid.*, p. 22), but taken up later for a detailed separate refutation (KS, pp. 102–11). The views Frege attacks have one, and only one, feature in common. Inductivism, physicalism, psychologism, and formalism are all different forms of empiricism.[2]

Though Frege feels certain that arithmetical truths are analytic and therefore *a priori,* he has to admit at the end of his book that he has made his claim only 'very probable.' He has shown that numbers are objects whose characterization involves concepts. He has not yet shown with absolute certainty that numbers can be defined as *logical* objects. How could one prove the existence of logical objects? Had Kant not shown that objects can be given only through sensibility and that their existence is consequently never assured by logic alone? In accordance with the Kantian conception the *Begriffsschrift* had dealt only with judgments and concepts. Where were logical objects to come from?

The *Foundations of Arithmetic* defines the numbers as extensions of concepts. It does not show that such extensions are logical objects. For that reason the book is not a self-contained work. It shows Frege's thought in transition from the stage represented by the *Begriffsschrift* to that represented by the first volume of the *Grundgesetze.* It shows him on the road towards the logical analysis of arithmetic, but not yet at the end of that road.

The question how numbers can be grasped as logical objects raises three issues to which Frege begins to address himself in the *Foundations of Arithmetic,* namely, (1) what is to be understood by a law of logic, (2) what is meant by an object, and (3) what the status of logical objects is. In the *Begriffsschrift* Frege had given almost no reasons why the formulas should be considered expressions of logical laws. From 1884 onwards he tries to explain logical laws as universal laws and as laws that explicate the notion of truth. The claim that numbers are objects Frege had first made in 1880 (cf. NS, p. 38). It leads to the question how there can be both objects given in sensibility and pure, logical objects. If there are logical objects, their existence cannot depend on the vagaries of the causal laws and they cannot be spatio-temporal. The problem then is to provide a characterization of the notion of object that allows for both spatio-temporal objects and pure objects. In the *Foundations of Arithmetic* Frege sets out to explain their status by calling them objective but not actual.

In order to understand the philosophical motivations behind the analysis of numbers offered in the *Foundations* we must first be clear about the over-all strategy Frege pursues with his analysis of arithmetic. We must then turn to an examination of the Fregean notions of logical law, logical object, and logical objectivity. Such an investigation takes one beyond the limits of the book itself. Only when these concepts have been clarified will Frege's procedure in the *Foundations of Arithmetic* be fully transparent.

On two different occasions Frege tried to explain the strategy that led him to consider the notions of logic, objectivity, and logical object. It seems useful to be clear about this strategy before we try to follow him in the examination of the three crucial notions. His reasoning is laid out in an uncompleted piece entitled '*Logik*' of 1897 (NS, pp. 155–61) and in the late essay 'The Thought' (KS, pp. 355–61).

On both occasions the unspoken assumption is the belief that

(1) Not all knowledge is empirical,

which Frege shares with both Kant and Lotze. The problem is to show that proposition (1) is true. Like Kant, he tries to do so by proving

(2) Mathematical truths, i.e. truths of arithmetic and geometry, are *a priori*.

Proposition (2) could be proved by establishing the *a priori* character of either geometry or arithmetic. There are two reasons why proof of the *a priori* character of arithmetical truth seems the preferable strategy to follow. For one thing, both Frege and Kant held geometrical truths to be based on *a priori* intuition. How was one to establish this fact beyond any possible doubt? It could always be disputed without falling into logical incoherence. Second, Frege believed that arithmetic is necessary for the justification of scientific induction. It is also necessary for the formulation of the more abstract empirical laws. To prove that arithmetical truths are *a priori* is therefore to prove not just that there are isolated pieces of *a priori* knowledge, but that *a priori* knowledge is fundamental to empirical knowledge.

The proof that arithmetical truths are *a priori* would be no more compelling than the corresponding proof for geometry if arithmetical propositions were merely synthetic. But if

(3) Arithmetical truths are laws of logic,

as Frege was inclined to believe with Lotze, then a definitive proof might be possible. What was needed for such a proof was a rich and precise logic in which the basic laws of arithmetic could be derived.

This seems to have been the point to which Frege had reasoned by the time he came to write the *Foundations of Arithmetic*. The question must then have risen in his mind: What is to count as a law of logic? As he reflected on this question, he knew that a widespread opinion held that

(4) Laws of logic are very general empirical laws of human reasoning.

It is this view that can be called 'psychologism.' If psychologism were true, the proof of (3) would hardly guarantee the truth of (2) and (1). Hence, it was not sufficient to construct a better logic and to derive arithmetic within it. The formal considerations needed to be supported by direct philosophical argument against psychologism.

At this point, the arguments of *'Logik'* and 'The Thought' become important. Frege was by no means unaware of the fact that the dominant form of psychologism in Germany was physiological in its orientation. It held that (4) is true because

(5) Human beings are natural, material beings whose properties can only be investigated empirically.

How was one to argue against this thesis? Frege's strategy was to try to show the incoherence of (5). The materialists were arguing in this way: Only what is immediately accessible to the senses exists. The empirical sciences tell us what is so accessible. The empirical sciences also tell us that the world is a material whole and that human beings are physiological mechanisms. Therefore, the truth of (5) is undeniable.

The question was whether that argument could stand up to scrutiny. How was empirical knowledge obtained? By observation, it seemed. Hence it seemed obvious that

(6) Only sensations or subjective ideas are immediately accessible to the senses.

If one assumed that only those things exist that are immediately accessible to the senses, the conclusion was clearly that

(7) Only subjective ideas exist.

The materialist, physiological view of human nature therefore seemed to collapse into subjective idealism or phenomenalism. By the time Frege was considering these points, scientific naturalism with its realistic materialism was in fact already giving way to this insight.

The next step was to show that subjective idealism was no more coherent than materialism. An idea must always be someone's idea. Had Kant not established this in his arguments against Hume? No subjective idea could exist without a subject and that subject in turn could not be an idea. Hence,

(8) There exists something which is not a subjective idea.

In the ideas which the subject entertains we can distinguish two components: that which makes the idea necessarily the possession of the subject to which it belongs and the thought expressed by the idea, which can in principle be entertained by more than one subject. The thought is not dependent on a subject. It is inter-subjective, self-contained, and complete in itself. The thought is objective, and it is the primary object which presents itself to the subject. It follows that

(9) Besides subjective ideas there is not only a subject, but there are also objective thoughts.

Such thoughts are not actual in the way in which the subject and his ideas are. If they were actual they would be in space and time and we would be able to interact causally with them. In that case our knowledge of such thoughts might be empirical. It is essential for the use to which Frege wants to put his thesis (that there are objective thoughts) that the knowledge of such thoughts cannot be empirical. Only if he can establish that point can he show that logic, in dealing with thoughts, is dealing with something non-empirical; that arithmetic, in so far as it is derivable from logic, is based on something non-empirical; and that therefore there is non-empirical knowledge which is the basis of empirical knowledge. It is essential, then, that thoughts not be actual. Their peculiar and irreducible status is described as that of being objective.

This consideration of Frege's strategy leads to the conclusion that his concern with the notions of logic and objectivity is deeply motivated by his epistemological position. The point deserves emphasis in the face of a prevailing view that ignores the epistemological aspects of Frege's thought and focuses instead on so-called ontological issues.[3] How little ontological issues concern him can be seen from the fact that much of what he says about objectivity can be read either as expressing a straightforward dogmatic ontology of the kind that is usually called Platonist or as developing a view that is Kantian in the sense that the objective may represent merely a necessary condition of understanding. It remains to be seen which of these two possibilities is realized in Frege's thought.

The question is significant also because of the peculiar twist Dummett has given to the discussion of Frege's ontology. Examining Frege's views he says: 'The attempt to interpret him otherwise than as a realist leads only to misunderstanding and confusion.'[4] In the passage in question Dummett unfortunately does not make explicit in which of the many possible senses he considers Frege a realist. It has already been argued that with respect to empirical objects Frege's views must be close to Kant's. Empirical objects are in space and time, but space and time are *a priori* forms of sensibility. That seems to imply that for Frege empirical objects can only be empirically real, but must be transcendentally ideal. That leaves open the question of the status of those non-spatio-temporal things which Frege calls objective, viz. thoughts, numbers, concepts, and so on. They cannot be said to be 'empirically real, but transcendentally ideal,' since their reality is not empirical and, if it is in some sense ideal, that is not because of the ideality of space and time.

Dummett's claim that Frege is a realist with respect to those things which he calls objective appears at times to mean that (a) Frege was a metaphysical realist with respect to empirical objects and (b) Frege considered objective things real in the same sense as empirical objects. Thus he writes in 1967:[5]

> Platonism, as a philosophy of mathematics, is founded on a simile: the comparison between the apprehension of mathematical truth to the perception of physical objects, and thus of mathematical reality to the physical universe. . . . This comparison strikes philosophers of an anti-metaphysical temperament as having the characteristic ring of philosophical superstition: but it cannot lightly be dismissed, because it has been endorsed by others as distinguished as Frege and Gödel.

What Dummett says here seems to fit better the views expressed by Gödel than those held by Frege. While Gödel believes in the analogy between visual perception and mathematical intuition, Frege neither takes recourse to the notion of intuition when he talks about our knowledge of arithmetic, nor does he assume that seeing an object and grasping a thought are equivalent activities (cf. KS, pp. 359ff, and NS, pp. 149–50).

In other passages Dummett gives a somewhat different account of what he thinks Frege's realism or Platonism amounts to. In 1963 he writes: 'As Kreisel has remarked, the issue concerning platonism relates, not to the existence of mathematical objects, but to the objectivity of mathematical statements.'[6] This judgment is repeated and elaborated in 1978:[7]

> The opposing views about the status of mathematical objects need play no part in resolving the critical disagreement, which relates to the kind of meaning that we succeed in conferring upon our mathematical statements. . . . Whether mathematical objects are mental constructions of ours or exist independently of our thought is a matter of what it is to which they owe their existence; whereas the important disagreement

between platonists and intuitionists is unaffected by this metaphysical question.

It is not immediately clear how this characterization of Platonism or realism is to be squared with the previous one. The ambiguity between the two pervades Dummett's whole discussion of Frege's views.[8] Nevertheless, it is clear from Dummett's writings that when he criticizes Frege's realism his predominant concern is with the objectivity of mathematical statements. What is initially an ontological question (Is Frege a realist or not?) is reinterpreted by him as an epistemological one (What kind of knowledge does Frege claim to have of mathematical truths?) and that in turn is understood as a logical question (What kinds of inference does Frege allow himself in mathematics? Is his reasoning in accord with two-valued classical logic, or is it reconcilable with intuitionistic assumptions?).

Contrary to Dummett's assumption we should not presuppose that everything that is important in the ontological and epistemological question is captured in the logical one. Metaphysical Platonism may indeed be an implausible doctrine, but it is not clear that intuitionism is the only or the most compelling way to avoid those implausibilities. Not only does intuitionist logic compel one to abandon modes of reasoning that seem innocuous and natural, but it creates such complexities in mathematics, particularly in real-number theory, that it has never gained the allegiance of a very large number of mathematicians. In addition, intuitionist logic has its own internal problems of plausibility.[9] Given the choice between the implausibilities of metaphysical Platonism and those of intuitionism, it is not at all evident that we should opt for the latter.

Kant's philosophy can be seen as an attempt to mediate between the ontological claims of realism and the epistemological claims of idealism. His resolution of the dispute consists in the thesis that knowledge is objective, but objective only for us, that the objective is not independent of reason. He thought that our classical modes of reasoning could thus be safeguarded without committing us to unwanted metaphysical conclusions. If Frege's theory of objectivity can be interpreted in this Kantian sense, we can credit him with an understanding of the shortcomings of metaphysical realism or Platonism while holding on to the belief in the objectivity of logic and mathematics. There is a sense in which that position can be called realism but its realism is not incompatible with idealism: it is itself a form of idealism.

The examination of the Fregean notions of logic, objectivity, and logical object, to which the text of the *Foundations of Arithmetic* gives rise, thus leads us to the deepest level of Frege's thought.

IN SEARCH OF LOGICAL OBJECTS

3 Logical laws as universal

In the *Begriffsschrift* Frege said very little about his understanding of the notion of logic beyond the fact that logic is concerned with necessities of thought rather than with factual matters or natural necessities (BS, p. xii). Three years later, he added to this the claim that logic deals with the forms of thought rather than with actual contents (*ibid.*, pp. 113–14) and that the principles of logic must be universal (*ibid.*). In making these assertions, he made no attempt to explain terms such as 'necessities of thought,' 'forms of thought,' and 'universality.' They were clearly borrowed from Kant and used without systematic reflection. Only at about the time of the *Foundations of Arithmetic* did he become aware of the need to clarify them.

Neither in the *Begriffsschrift* nor in the *Foundations* does he try to explain logical truths in the terms with which we are most familiar from Leibniz and Kant. In the light of Frege's familiarity with Leibniz's works it is striking that he never even mentions the Leibnizian characterization of logical truths as truths in all possible worlds. We can only speculate that it was his assumption that the concepts of necessity and possibility are not legitimate logical notions at all which stopped him from pursuing the Leibnizian characterization (cf. BS, pp. 4–5). In the *Foundations of Arithmetic* Frege introduces the Kantian notion of analyticity but he does not characterize logical truths as analytic, because in his improved definition of the term an analytic truth is one derivable from logical laws and definitions alone (F, p. 4), so the notion of analyticity presupposes the notion of logical truth.

In the *Foundations of Arithmetic* itself Frege has almost nothing to add to his earlier informal characterization of the notion of logical truth. What matters to him here, and is equally important in his later writings, is the idea that logical laws are strictly universal. Propositions are analytic, he says in the *Foundations*, if one can prove them by 'making use only of truths which are of a universal logical nature and without those which belong to the sphere of some special science' (*ibid.*).[10] Later he repeats that logic deals only with 'the most universal which has validity for all areas of thinking' (NS, p. 139). And he goes on to say that 'logic is the science that deals with the most universal laws of truth' (*ibid.*). In the *Foundations of Arithmetic* he says that there must be strictly universal laws if there are to be any general laws at all. 'If we recognize the existence of general truths at all we must also admit the existence of such primitive laws, since from mere individual facts nothing follows, unless it be on the strength of a law' (F, p. 4).

The idea that logic deals with strictly universal laws and is therefore applicable to everything thinkable is for Frege the basis of the claim that arithmetical truths are logical laws. In an essay written one year after the *Foundations of Arithmetic* he distinguishes two senses in which one might say that arithmetic is formal. In the first sense, we regard arithmetical equations as uninterpreted formulas. According to this view, 'arithmetical equations have as little sense as an arrangement of figures on the chessboard expresses a truth' (KS, p. 109). Frege tries to show that this conception is ultimately incoherent. The formalist mathematician will at least want to assure himself of the consistency of the rules by which we operate in the uninterpreted arithmetical calculus. But Frege holds that 'it is most improbable that a strict proof of the consistency of the rules of calculation can be given without leaving the ground of the formal theory. And, even if that should be possible, it would be insufficient, since that which is consistent is not necessarily true' (*ibid.*, p. 110). He raises the same objections in many other places.

Formalist arithmetic, he tries to show, is really an incoherent undertaking. Its apparent plausibility derives from confusing it with the formal or logical conception of arithmetic. Almost anything can be counted that is an object of thought, 'the ideal as well as the real, concepts as well as things, the temporal as well as the spatial, events as well as bodies, methods as well as theorems' (*ibid.*, p. 103). Laws that deal with such a universal subject matter 'must rightly be reckoned to logic. . . . There is no sharp boundary between logic and arithmetic. . . . If this formal theory is correct, then logic is not as fruitless as it may superficially appear. . . . This science is capable of no less precision than mathematics itself' (*ibid.*, pp. 103–4). Arithmetic, as well as logic, deals with everything that is thinkable. And thinking is a unitary phenomenon. 'Thought is essentially the same everywhere; there are no different kinds of laws of thought for different objects' (F, p. iii). Though mathematical induction is sometimes considered a form of reasoning peculiar to mathematics, a kind of aggregative thinking, Frege intends to show that such induction too 'is based on the universal laws of logic and that one does not need special laws of aggregative thinking' (*ibid.*, p. iv).

The conception of arithmetic as consisting of strictly universal laws and therefore as part of logic brings with it certain difficulties which seem to demand further clarification of the concept of the logical. Among the propositions of arithmetic are not only those that make claims about all numbers, but also those that make assertions about particular numbers and others again that assert the existence of numbers. The question is how such propositions could be regarded as universal, and therefore logical, truths. A universal truth, Frege says, contains no reference to a particular object (*ibid.*, p. 4), but it seems that some arithmetical propositions contain such reference. This observation leads Frege to draw two very important conclusions.

The first is that all existential propositions of arithmetic must be derivable from non-existential, universal laws if the claim that arithmetic is logic is to be upheld. If we look at the formal system of the *Grundgesetze* we discover that in fact none of the six basic laws or axioms is existential in form. The introduction of value-ranges is not justified by axioms which say that under such and such conditions there exists a value-range such that . . . , but by an axiom expressing a universal truth concerning first-level functions and their corresponding value-ranges. Frege can avoid existential axioms in logic because of the particular character of the axiom V of the *Grundgesetze* through which value-ranges are introduced into the theory. This axiom allows one to derive existential conclusions about value-ranges and, since numbers are introduced as value-ranges, existential conclusions about numbers. It is axiom V which permits Frege to hold both that logical laws are strictly universal and that particular and existential arithmetical propositions can be derived from them. Unfortunately, it is precisely this feature of the axiom that makes it possible to derive a contradiction in the system of the *Grundgesetze*, as Russell was the first to recognize.

In order to salvage the possibility of the derivation of arithmetic from logic Russell therefore found it necessary to introduce an explicitly existential axiom into his theory of types, i.e., the so-called axiom of infinity. Russell's own attitude towards the axiom was ambiguous. On the one hand, he was convinced of the reducibility of arithmetic to logic and of the need for something like the axiom of infinity for that reduction. On the other, he was clear that this axiom could not be derived from more fundamental assumptions.[11] His conclusion is therefore that 'the axiom of infinity will be true in some possible worlds and false in others; whether it is true or false in this world, we cannot tell' (*ibid.*, p. 141). Given that its truth depends in this way on the character of the world that happens to exist, the axiom seems to have lost its status as a logical truth. In order to overcome these difficulties, F. P. Ramsey devised an ingenious but doubtful argument to show that the axiom of infinity must be a tautology or logical truth after all.[12]

Whatever we think of Ramsey's argument, it is clear that it involves a concept of logic that is substantially different from Frege's. Given Frege's Kantian notion, all logical truths must be ultimately universal; whereas on Ramsey's conception there can be irreducibly existential laws of logic. The thesis that arithmetical laws are reducible to existential logical principles might be called *ontological* logicism, whereas Frege's logicism is supposed to be non-ontological in character. The discovery of Russell's antinomy has shown that non-ontological logicism is impossible. All later forms of logicism have therefore been ontological. But for Frege ontological logicism is an impossibility. Once he recognized the impossibility of his own program he therefore abandoned logicism altogether. To some that has seemed an extreme overreaction, but it was unavoidable given Frege's understanding of what logic is.

The Fregean claim that logical principles must ultimately be universal, conjoined with the observation that arithmetical propositions are not always universal in form, led Frege to another significant conclusion. He saw that he needed to explicate the notion of universality that had been in his mind. The laws of logic and arithmetic are universal even if their external syntactical form is not that of a universal proposition. There is another and deeper sense in which such laws can be called universal. They are universal because they are universally applicable.

How was one to understand this notion of universal applicability? At around the time of the writing of *Foundations*, Frege was led to the notion of truth. All human understanding and all science aims at truth. It follows that laws that deal with the notion of truth must possess universal significance. If we understand logical laws as being concerned with the concept of truth in a special and unique way, then we can explain the universal applicability of the logic laws. But the difficulty is to show how logic is concerned with truth.

In all of Frege's writings from the time of the *Foundations of Arithmetic* onwards the notion of truth plays a prominent and indispensable role. Around 1884 he went so far as to say that 'someone who does not grasp the unique meaning of this word, may also not be clear about the task of logic' (NS, p. 3). This is somewhat surprising when we remember that the *Begriffsschrift* of five years earlier had not even mentioned the notion of truth. He had not even spoken there of judgeable contents as true or false, but had simply distinguished them into the affirmed and the denied. In 1880/81 that remained his preferred terminology (cf. NS, p. 40). By 1884, however, the situation has changed. In the *Foundations of Arithmetic* he speaks explicitly of truth and falsity. This change in terminology and the new awareness of the significance of the notion of truth for the understanding of logic is accompanied by a new interest in the notion of objectivity.

The connection between the notion of truth and that of objectivity is emphasized in another text which Frege must have written around 1884. It is the first of a series of attempts to present his logic in an informal manner without the use of his unfamiliar symbols (*ibid.*, pp. 1–8).[13] Frege made several attempts at such an informal presentation. The first three remained in fragmentary form. The second was undertaken around 1897 (*ibid.*, pp. 137–63); the third exists in two slightly different versions written in 1906 (*ibid.*, pp. 201–18). The fourth and final attempt at such a presentation appeared in the form of the essays 'The Thought,' 'Negation,' and 'Thought Connection' in the years between 1918 and 1923 (KS, pp. 342–94).[14] All these projects have a roughly similar structure. All begin with a discussion of the notion of truth and move from there to the examination of the notion of objectivity. My conjecture is that Frege undertook these projects not only to make his ideas accessible to those who found the symbolism impenetrable, but also as an essential part of the attempt to clarify the notion of truth.[15]

IN SEARCH OF LOGICAL OBJECTS

4 Logical laws as laws of truth

We must now look more carefully at the Fregean notion of truth as the key to an understanding of his notions of logic and objectivity.

In the fragment called *'Logik'* of around 1884 the notion of truth is approached through a discussion of the difference between logic and psychology. Frege writes:

> In natural thinking the psychological and the logical have become intertwined. It is our task to isolate the purely logical. That does not mean to ban the psychological from actual thinking, which would be impossible; but rather to consider only the problem of logical justification. The postulated separation of the logical from the psychological therefore consists only in their conscious distinction (NS, p. 6).

In later writings Frege insists that psychology and logic have entirely different subject matters. The former deals with ideas, which are subjective mental phenomena, whereas the latter is concerned with objective thought. In *'Logik'* the distinction is drawn in a slightly different way.

Both psychology and logic can be said to be concerned with beliefs and judgments, but they are concerned with them in two different ways. Psychology investigates the cause of our beliefs, logic and epistemology investigate their justifying reasons. Frege asserts that all our judgments are causally determined but that not all such causes are justifying reasons. Empiricism, he argues, ignores this distinction and because of the empirical determination of our thinking regards all knowledge as empirical (*ibid.*, p. 2). But to describe how human beings have come to consider something as true is not the same as to give a proof of it. 'In science the history of the discovery of a mathematical or natural law cannot replace its rational justification' (*ibid.*, p. 3). Some beliefs are justified on the basis of previously recognized truths. Such justification is called inference and logic is concerned with stating the laws of right inference. But there must also be judgments which either require no justification at all or whose justification does not rest on other judgments. Epistemology is concerned with justifications of a non-inferential sort (*ibid.*).

The crucial notion which distinguishes logic and epistemology from psychology is the notion of truth. Justifying reasons always have to do with the truth of a judgment. 'Causes which determine acts of judgment do this according to psychological laws; they may lead to error as much as to truth; they have no internal relation to truth; they treat the opposition of true and false as indifferent' (*ibid.*, p. 2). Psychology is, of course, also concerned with truth, but not with the truth of the judgments whose causes it investigates empirically. The goal of all scientific activity is truth (*ibid.*). The question is in what specific sense logic is concerned with that notion. Frege writes: 'It would perhaps not be wrong to say that the laws of logic are nothing but the development of the content of the word "true" ' (*ibid.*, p. 3).

Logic is thus distinguished from both psychology and epistemology. Its concern is with the notion of truth, and specifically with the notion of valid inference, that is, with the justification of the truth of one proposition on the basis of the truth of others. The laws of right inference are the *Gegenstände* of logic, i.e., its objects (or, more idiomatically, its subject matter). These objects are independent of our recognition and, in that sense, complete in themselves, and they can be exactly the same for different human beings. Completeness and identity, the criteria which Frege associates in the *Foundations of Arithmetic* with objects such as the numbers, are here associated with the logical laws, which are considered the proper objects of logic. Moreover these objects are considered to be objective (*ibid.*, p. 7).

The interconnection between the notions of truth, object, and objectivity is reaffirmed in the later projects. In the second of the four attempted informal presentations of his logic Frege enriches his characterization with one important addition. He introduces the notion of a law of truth and does so with an apology: 'Logic is the science of the most general laws of truth. One might consider it impossible to find anything definite expressed by that claim. The awkwardness of the author and of language may be to blame for this' (NS, p. 139). In spite of this hesitation the term remains part of Frege's terminology. In fact, the essay 'The Thought' begins by reasserting the claim of 1897.

> The word 'true' sets the goal for logic, just as the word 'beautiful' does for aesthetics and 'good' does for ethics. Admittedly, all sciences aim at truth, but logic is also concerned with it in quite another way. It relates to truth in roughly the way physics relates to weight or heat. To discover truths is the task of all sciences; it is the concern of logic to recognize the laws of truth (KS, p. 342).

In both places the characterization is elaborated with the comment that the meaning of the word 'true' is basic and simple (NS, p. 137; KS, p. 344). That means that no definition of the term is possible. Since Tarski we have become used to thinking that in certain formal languages the notion of truth is definable. Frege's point is not to deny this formal possibility, but to maintain that the terms in which such a definition would have to be cast would always be less fundamental than the notion of truth itself. Thus, we might explain the truth of the sentence 'F(a)' by saying that it is true if and only if the object a falls under F, has the property F, or satisfies F; those explanations involve notions such as object, falling under, having a property, or satisfying a predicate. On Frege's account each of them is less basic than truth itself; it is the most fundamental logical notion for him.

IN SEARCH OF LOGICAL OBJECTS

Since truth cannot be defined, in particular it cannot be defined as correspondence between propositions and facts. Kant had made a similar point in the *Critique of Pure Reason*. 'A sufficient and at the same time general criterion of truth cannot possibly be given' (B 83), he said, and therefore the correspondence theory is valid only in a trivial sense. In Kantian thought facts are given only in judgments and hence there are no judgment-independent facts with which our judgments could be compared. Frege's reasons for rejecting the correspondence theory of truth are similar. His argument against it is borrowed from the objections Lotze raises in his *Logik*. For Lotze the failure of the correspondence theory is what explains the persistence of philosophical scepticism. Truth is initially defined as that which corresponds; then it is noticed that correspondence between an idea and an independent fact is impossible, and the sceptical conclusion is that truth is unknowable. 'It is never anything but the connection of our representations that constitutes the subject matter of our investigations' (*Logik*, p. 491). He thinks the correspondence theory fails for a simple reason: If the truth of P is to be tested by Q, then that amounts to testing the truth of P by the truth that Q is the case. In order to establish that something corresponds to the facts we must know whether it is true that it so corresponds (*ibid*., p. 492). Frege repeats these reflections almost literally (NS, pp. 139–40; KS, p. 344). The failure of the correspondence theory is for him evidence that no definition of truth can succeed. Every explanation must fail because it would be of the form: 'A is true if it has such and such a property or stands in this or that relation to such and such a thing.' No such explanation can succeed because 'in every case as in the given one everything would depend on whether it is true that A has such and such a property or stood in this or that relation to such and such. Truth is obviously something so basic and simple that the reduction to something even simpler is impossible' (NS, p. 140).

Given that logical laws cannot provide a definition of truth, the question is in what sense they are to be understood as laws of truth. We might still think it means that the notion of truth must somehow occur explicitly in the logical laws. Since the 1930s logicians have come to think more and more that semantics is the real core of their discipline. One might think that Frege's characterization of logic is meant to say something very similar; but in fact that is not the case. Semantic considerations are for him primarily designed to facilitate the construction of the logical symbolism; they have only an auxiliary function.[16] That makes Frege's characterization of logical laws as laws of truth even more puzzling.

To clarify the point it is necessary to consider exactly how the notion of truth enters into Frege's logical constructions. When we look at his symbolic language we notice that it contains no symbol for truth or falsity. Those notions enter into Fregean logic only through true and false sentences that can be constructed in the symbolic notation. But the absence of symbols for the terms 'true' and 'false' is no omission. Frege believes that there is no need at all for such terms in his notation. To say that '*A* is true' is to say exactly the same as '*A*'; the word 'true' is really redundant. Wherever it occurs it can be eliminated, though that may involve certain grammatical transformations. Frege writes: 'The word "true" has a sense that contributes nothing to the sense of the whole sentence in which it occurs as a predicate' (NS, p. 272). The reason is not that the word has no meaning, but rather that 'it is distinguished from all other predicates by always being asserted when anything is asserted' (*ibid.*, p. 140). For 'it is really the form of the declarative sentence through which we assert truth and we do not require the word "true"' (*ibid.*).

The notions of truth and falsity have a function in our language, not because the words 'true' and 'false' occur in it, but because we can grasp thoughts and can come to recognize them as true or false in judgment and assertion. Thoughts are true or false independently of our acts of recognition, but for us the objective content with which logic deals can only be grasped in subjective acts of judging and asserting. It is in those acts that truth and falsity have their place for us. The laws of logic are in one sense laws of truth because they give us justifications for our assertions.

> One justifies a judgment either by going back to truths already recognized or without using other judgments. Only the first case, that of inferring, is the concern of logic. The examination of concepts and judgments is only a preparation for the examination of inferences. The task of logic is the establishing of laws according to which one judgment is justified by others regardless of whether they are true or not (NS, p. 190).

What is surprising in Frege's characterization of logic in terms of inference and thereby in terms of judgment is the fact that he has thus introduced a notion which he himself regards as psychological. The dilemma is serious. One can ask about the nature of the relation that obtains between a thought and its truth. It is a relation that obtains quite apart from human acts of judging and asserting, because the truth of a thought is objective. Truth is discovered and not invented. But we cannot *know* whether this relation obtains apart from making an act of judgment or assertion. The truth of a thought is given to us only in so far as we connect the thought with truth in a judgment. When we explicate the laws of truth we can do so only in judgments and assertions. What we can spell out is therefore the subjective relation that we establish between a thought and truth. It is in this sense that Frege writes:

> Thus the word 'true' seems to make possible the impossible, namely, to make that which corresponds to the assertive force appear to be contri-

buting to the thought. And this attempt, though it fails—or, more correctly, because it fails—points to what is peculiar to logic (NS, p. 272).

There is no way in which we can *describe* the objective relation that obtains between a thought and the truth. If we could, we would be able to form a thought expressing that relation and in it the word 'true' would make a contribution to the thought itself. For us truth is realized only in judgment and assertion; there is no other access to it. And for that reason the laws of truth with which logic deals are the laws justifying our assertions.

Truth is embedded in our language through judgment and assertion. In so far as we regard logic as dealing *explicitly* with the notion of truth, there can really be no such thing. How then does it happen that the word 'true,' despite the fact that it seems without content, also appears indispensable? Should it not be possible to avoid the word, at least when laying the foundations of logic, since it seems to create nothing but confusion? That we cannot do without it reveals that our language is imperfect. If we had a logically perfect language we might not need any logic or we might be able to read it off the language. But we are a long way from that goal. Work in logic is to a great extent a fight against the logical blemishes of language (*ibid.*).

If logic is meant to be a theory of meaning or a theory of truth, then there can really be no such thing. The appeal to meaning and truth is necessary only in so far as we are concerned with a critique of ordinary language or with the setting up of a logically perfect language. Once we have such a language, our semantic considerations will fall away. Frege's thoughts are not unlike those we find in Wittgenstein's *Tractatus*. There, too, the possibility of a semantic meta-theory for our language is rejected. As Wittgenstein puts it: 'Logic must take care of itself' (5.473).

The idea that for Frege semantic notions such as the notion of truth are merely auxiliary devices for the setting up of a logically perfect language can also be confirmed in a way that draws attention to another sense in which for him the laws of logic are laws of truth. How does Frege actually make use of the notions of truth and falsity? At one point he says that there are certain concepts characteristic of logic:

> Logic is by no means completely formal. If it were, it would be without content. As the concept of point belongs to geometry, so logic also has its own concepts and relations, and only because of that has it got any content. With respect to its own things logic is not formal. . . .
> To logic there belongs, for instance, negation, identity, subsumption, subordination of concepts (KS, p. 322).

It is striking that the notions of truth and falsity are not listed here among the characteristic concepts of logic. That is no omission. In the passage quoted Frege is talking about those notions for which his symbolism provides notations. Truth and falsity are not among them.

Nevertheless, Frege does not say that logical laws explicate notions such as negation, identity, and so on, but that they are laws of truth and explicate the notion of truth. One can understand this claim better if one considers how symbolic notation is introduced into the logic. In the *Grundgesetze* each symbol is described syntactically and its meaning is explained by indicating under what conditions a sentence containing the symbol is true or false. We could say that for Frege logical laws are true in virtue of the logical terms occurring in them and the meaning of those terms is explicated with reference to the notions of truth and falsity. It is in this sense that the notion of truth seems to be necessary for the design of a logically perfect language. Once the terms of that language have their meaning and the language is properly set up, no explicit reference to truth and falsity will be necessary any longer.

5 *Frege's Lotzean notion of objectivity*

Truth is objective. That means that propositions or thoughts are true or false quite independent of our ability to assess their truth-value. When Dummett characterizes Frege as a realist or Platonist it is this aspect of the notion of objectivity that he focuses on. But the notion of objectivity has a much wider application. Not only is the truth of a thought objective, but the thoughts themselves are also objective and so are numbers, value-ranges, concepts, relations, and many other things.

The notion of objectivity is essential to Frege's refutation of psychologism and thus to his anti-empiricism. It has also given rise to arguments over whether and in what sense Frege is concerned with ontological issues and whether and in what sense he is a realist. In order to get clearer this crucial notion one must take seriously the three things Frege says about it.

(1) The objective is that which can be grasped by more than one human (rational) being. The objective, in other words, is the intersubjective.

Thus, he holds that a thought is 'not the product of an inner process, but something objective, i.e., something which is exactly the same for all rational beings who can grasp it' (NS, p. 7). Mankind has a common store of thoughts; many people can collaborate in the development of one science.

(2) The objective is that which does not require a bearer.

Ideas are subjective in the sense that we can legitimately only speak of this or that person's ideas; and, because of this essential subjectivity, they are not really communicable. On the other hand: 'We are not bearers of the thought, as we are bearers of our ideas. . . . When we are thinking we do not create a thought, but we grasp it. . . . What I recognize as true I judge to be independent of my recognition of its truth' (KS, pp. 358–9).

(3) The objective must be distinguished from that which is *wirklich*, i.e., actual or real.[17]

Frege writes: 'I distinguish the objective from the tangible, the spatial, the actual [or real?]. The axis of the earth, the center of mass of the solar system, are objective, but I would not call them actual [real?] as is the earth itself' (F, p. 35).

Some insight into how the notion of objectivity is to be understood can be derived from the fact that Frege took it from Lotze's *Logik*. In that work Lotze says explicitly that objectivity 'does not in general coincide with the *Wirklichkeit* that belongs to things' (*Logik*, p. 16). And he maintains that the objective is that 'which is the same for all thinking beings and which is independent of them' (*ibid.*). In other words, he endorses all three of Frege's assertions about the objective.

At the same time Lotze makes clear that the doctrine of objectivity is not to be taken as ontological, but rather as epistemological. The issue is discussed in the third book of the *Logik*, which deals specifically with problems of knowledge. It begins with an examination of scepticism, which Lotze argues is ultimately incoherent because 'it cannot get rid of the assumption of truth which is valid in itself' (*ibid.*, p. 497). Scepticism derives its initial plausibility from its observation that the commonly held view of truth is untenable. The common view is that 'our cognition is made to picture a world of facts,' that truth consists in the correspondence 'of our cognitive picture with states of affairs' (*ibid.*, p. 490). For the reasons already given, the correspondence theory must be rejected. But there is no reason to abandon the notion of truth altogether. 'The confidence that truth can be found through thinking is the inevitable precondition for all investigating' (*ibid.*, p. 492).

Lotze next reformulates scepticism in terms of the Heraclitean or pseudo-Heraclitean theory of flux that is attacked in Plato's *Theaetetus* (*ibid.*, pp. 504–5). He credits Plato with recognition of the fact that the senses may present us with changing impressions but cannot give us a stable external world. For that we require in addition something else, namely, 'a pervasive truth' (*ibid.*, p. 508). Our knowledge of changeable objects presupposes an 'unchangeable system of thoughts' (*ibid.*).

But how is the Platonic theory to be understood? What kind of being do Platonic ideas possess when they are not actually entertained by a human mind?

> In so far as we have and grasp ideas, they possess reality [*Wirklichkeit*] as events, as things that happen in us . . . but their content, considered separate from the mental activity we direct towards it, is not something that happens. It does not exist in the way in which things exist; it is simply *valid*. We cannot ask what validity is on the assumption that an understanding of it could be derived from something else. . . . Just as one cannot understand how it comes about that something exists or happens so one cannot explain how it happens that a truth is valid. One must regard the notion as fundamental and as resting only on itself (*ibid.*, pp. 512–13).

Lotze regards the ontological interpretation of Plato's ideas according to which they have a reality separate from but similar to that of things as an 'outlandish claim' (*ibid.*, p. 513). Plato's assigning the ideas to a non-spatial heaven was not an attempt 'to hypostasize their mere validity into a kind of real existence, but the noticeable effort to ward off any such interpretation' (*ibid.*, p. 516). Nevertheless, the lack of an adequate terminology makes it difficult for him to say what he really wanted to assert, and out of this difficulty have come the false ontological interpretations and the useless quarrels over the validity of these interpretations of the Platonic doctrine.

> Though Plato asserted only the eternal validity of the ideas, but not their real existence [*Sein*], to the question of what they were he could do nothing but bring them under the general term οὐσία [being]. And thus the door was opened to a misunderstanding that has gone on till today (*ibid.*, p. 516).

The misunderstanding is due to a specific shortcoming in Plato's account. The notion of validity applies, in the first instance, to propositions; it can be applied to concepts only indirectly. 'Of them we can say that they have meaning, but only because propositions are true of them' (*ibid.*, p. 521). The ontological misinterpretation of the Platonic doctrine has its roots in the separation of concepts from their propositional contexts.

The question here is not whether Lotze's interpretation of the Platonic theory is correct or not, but only how Lotze understands that theory. He takes it as an epistemological, rather than an ontological, theory, and therefore he is an epistemological rather than an ontological Platonist. That is to say he believes with Plato that empirical knowledge of temporal, changing things presupposes some knowledge of non-temporal, non-changing things. On this view, the timeless and non-historical is conceived as the foundation of the temporal and the historical.

In accordance with medieval usage the doctrine of ideas interpreted ontologically is often called a realistic doctrine. It is in this sense that Frege is usually considered a realist. But, given the close connections between Frege and Lotze, we may wonder whether this is the correct interpretation. There are certainly Platonic elements in Frege's thought, but they might not be the ones interpreters have been looking for. Is it not possible that Frege, like Lotze, was an epistemological, and not an ontological, Platonist?

The parallels between Frege's and Lotze's thinking make it plausible to assume that their arguments aim in the same direction. Frege agrees with Lotze that the correspondence theory of truth fails. Truth and Falsity are not definable, they are the presuppositions of all our thinking. The True and the False 'are recognized, even if silently, by everyone who makes judgments at all, hence also by the sceptic' (KS, p. 149). Thoughts are objective precisely because their truth is independent of our recognition of it (ibid., p. 359). This objective realm guarantees the stability of the external world. 'If everything were in continual flux, and nothing maintained itself for all time, there would no longer be any possibility of getting to know anything about the world and everything would be plunged into confusion' (F, p. vii). Objectivity applies, first of all, to complete thoughts. Mistaken ontological explanations of numbers as properties of physical things, as subjective ideas, or as agglomerations of things have their source in the separation of the number term from the propositional context.

On the basis of these affinities it seems plausible to hold that Frege's doctrine of objectivity, like Lotze's, was intended as an epistemological thesis and that he was a critical rather than a dogmatic thinker. Three crucial passages lend direct support to this reading, two of them occurring in the *Foundations of Arithmetic*, and the third in the late essay 'The Thought.' In the first passage Frege writes:

> By objectivity I mean independence from our sensing, intuiting, and imagining and from the construction of internal pictures from memories of earlier sensations, but not independence from Reason; to say what things are like independent of Reason would be to judge without judging, to wash the fur without wetting it (F, p. 36).

The point seems to be that it is incoherent for us to try to say what things are in themselves, independent of our judgments, and that the claim that something is objective is not a dogmatic metaphysical claim.

In the second passage Frege is concerned with the real object (*der eigentliche Gegenstand*) of arithmetic, which, according to him, is Reason itself.[18]

> We are concerned in arithmetic with objects that come to be known to us not through the mediation of the senses, as something alien, from outside, but with objects that are immediately given to Reason and which Reason can see through completely as its most characteristic possessions (F, p. 115).

Interpreters of Frege's writings have occasionally noticed the Kantian slant of these remarks, but have insisted that in his later years Frege became increasingly concerned with the defence of a (dogmatic) metaphysics of a Platonistic kind.[19] The essay 'The Thought' is usually referred to in support of this reading, but even there Frege does not repudiate what he had said in the *Foundations of Arithmetic*. He writes:

> Neither logic nor mathematics has the task of investigating souls and the content of consciousness whose bearer is an individual human being. Rather, one might perhaps set down as their task the investigation of the Mind—of the Mind, not of minds (KS, p. 359).

Frege notes that to say we grasp objective thoughts is to speak in a metaphor. 'What I hold in my hand can be considered the content of my hand, but it is the content of my hand in quite a different sense and is more alien to it than the bones, the muscles of which it consists, and their tension' (*ibid.*). If we take this analogy seriously, it seems to imply that Frege does not hold that thoughts are in the mind as the bird is in the hand, but rather as the muscles and bones are in the hand. The objective is not something alien or external to the mind, but constitutive of it. It is its most characteristic possession.[20]

6 *Frege's concept of an object*

Having discussed the notions of logic and objectivity, we can now turn to the notions of object and logical object.

Frege's understanding of the notion of an object must be contrasted with the common view according to which the term 'object' stands for an empirical notion acquired through learning by ostension. This view holds that (1) there are objects in the world; (2) these can be known through acquaintance; (3) acquaintance with objects is achieved through straightforward sensory perception; and (4) the term 'object' is learnt when someone points to an object with which we are acquainted and says something like 'This is an object.' On this view, the paradigmatic object is a spatio-temporal entity of a size and location that allow for immediate sensory perception of it.

This common-sense view of objects has been criticized repeatedly by philosophers, particularly those inclined towards idealism. Kant, for instance, denied outright that there are things apart from the appearances and that things as appearances are known by simple acquaintance. For him the notion of an object is a formal, rather than an empirical, one. He agrees with the common-sense view only in assuming that sensibility is required for knowledge of objects. But, on his conception, sensibility alone is not sufficient.

Frege's view of objects is akin to Kant's in that it also considers the notion of object a formal notion, but it diverges from it by denying that sensibility is necessary for knowledge of objects (cf. F, p. 101). One reason for this disagreement is that for him, as for Lotze, the content of a judgment, i.e., the thought, is the paradigmatic object. But thoughts are not spatio-temporal, and so are not given to the senses.

Frege characterizes objects by the properties of the expressions that stand for them. In other words, he transforms what looks like a material and ontological problem into a formal semantic one. An expression standing for an object is called a proper name by Frege. That terminology is perhaps not altogether happy, since it turns out that the term has an essentially wider use for Frege than it does in everyday life. But nothing hangs on Frege's usage. Instead of using the two terms 'object' and 'proper name' he could have used two newly coined terms A and B and expressed his views completely in terms of them. All that Frege's doctrine requires is that what we usually call an object should fall under A and what we usually call a proper name fall under B. But in talking about Frege's text it is advisable to follow his actual usage.

In the narrowest (and perhaps most common) sense a proper name is an expression such as 'Theodore Roosevelt.' In a somewhat wider use of the term we call expressions like 'The Bank of America' or 'Her Majesty's Government' proper names. Even wider is the use of the term when it means any expression referring to an individual object. In this third sense 'the former capital of Germany' is a proper name of Berlin. A proper name in this sense is what is sometimes called in philosophy a 'referring expression.' Frege occasionally uses the term 'proper name' in this sense. But more commonly he uses the term in a wider sense. Frege offers four criteria for recognizing proper names:

(1) The expression does not begin with an indefinite article. (A definite article at the beginning, on the other hand, *may* be a sign for a proper name.)
(2) The expressions contains no free variables.
(3) The expression cannot properly occur as a predicate in a sentence. (It may, on the other hand, be part of a predicate.)
(4) The expression can occur on the left and the right of an identity sign and thereby form a complete sentence.

On Frege's understanding of these criteria it turns out that sentences are proper names. They clearly fulfill the second and third conditions. In Frege's symbolism they also satisfy the fourth, since he makes no distinction between the notions of identity and material equivalence of sentences. Sentences are for him therefore proper names and what they stand for are objects. The formal criteria for proper names embody the claim that objects are complete and carry identity conditions with them.

On the basis of these criteria Frege concludes that number terms are proper names and that the numbers they stand for are objects. Expressions like 'the number five,' 'the positive square root of two' begin not with indefinite, but with definite, articles. The expressions contain no free variables. They cannot be used predicatively, though they may be parts of predicates such as 'is identical with the number five.' And the expressions can be used to form sentences when combined with the identity sign.

In the assertion that number signs are proper names there is hidden the implicit assumption that these signs name something. That is an assumption which has been disputed by mathematical formalists who simply identify the numbers with the number signs and treat arithmetical equations as uninterpreted sign systems. Frege, of course, completely rejected mathematical formalism. He believed that the formalist doctrine could not even be stated coherently and that formalism in the end was always reduced to an interpreted arithmetic. In Frege's eyes the formalists lacked an adequate account of meaning, failing to recognize that mathematical expressions are meaningful only in so far as they stand for something.

7 *The logical analysis of natural numbers*

Instead of pursuing the issues of formalism or the proper explication of the notion of meaning, I turn now to Frege's claim that numbers are not just objects, but logical objects. This claim was fundamental to Frege's thinking. Years later he could write: 'As the basic problem of arithmetic we can consider the question: how do we grasp logical objects, in particular the numbers? Through what are we justified in regarding numbers as objects?' (GG, vol. 2, p. 265).

The question was most acute for someone brought up in the context of Kantian thought, since Kant completely rejected the notion of a logical object. Though Frege parts company with him on that point, his approach nevertheless remains influenced by Kantian considerations.

In the *Begriffsschrift* Frege had asserted the (Kantian) thesis of the priority of judgments over concepts. That assumption guided him in the very construction of his logic. Later reflection on the notion of a logical object took him back to these considerations. Kant had argued that there are pure concepts of the understanding — among them negation, implication, universality, and existence — which are not derived from intuition, but are given by the forms of judgment. Frege agreed with Kant's claim that judgments possess an original unity and he reasoned, with Kant, that if there were logical objects they would have to be defined in terms of that pure formal unity. Empirical, psychological, and formalist conceptions of numbers have their origin in the separation of the number term from the propositional context. The characterization of numbers as logical objects had to proceed in another way. Since logical objects cannot be given in

IN SEARCH OF LOGICAL OBJECTS

intuition, the meaning of the terms referring to them must be completely determined by the sentential contexts in which they occur. These considerations led Frege to emphasize again and again in the *Foundations* the importance of the context principle of meaning.

> That we can form no idea of the content of a word is . . . no reason for denying all meaning to it or for excluding it from our vocabulary. We are forced to the opposite conclusion because when asking for the meaning of a word, we insist on considering it in isolation, and that leads us to accept an idea as the meaning. Accordingly, any word for which we can find no corresponding mental image appears to have no content. But we ought always to keep before our eyes a complete proposition (F, p. 71).

In order to grasp the meaning of number terms we must consider the kinds of sentences in which they occur and the various uses to which those terms can be put in those sentences. Number terms typically occur in arithmetical equations; but it is equally important that they occur in other types of sentences. We say not only '$7 + 5 = 12$,' but also 'I saw two twins three days ago.' The philosophically most important question is to ask how number terms are used in sentences of the second kind. 'We can surely demand of arithmetic that it provide us with the means for understanding any application of number even if it is not itself concerned with such application' (*ibid.*, p. 26). Frege concludes from his examination of the non-arithmetical propositions in which number terms occur that the basic use of the natural numbers is for counting. Any acceptable definition of the natural numbers must take into account the fact that they are counting numbers.

Before we can proceed to the definition of the natural numbers we must therefore ask what counting is. This investigation gives rise to an important observation. It is that in order to count anything we must have a counting unit.

The same external phenomenon, Frege argues, can be counted in more than one way. Homer's *Iliad* can be called one poem, twenty-four books, or a large number of verses (F, p. 28). Given a pack of playing cards I cannot say which number attaches to it unless I am given a further word, such as 'card' or 'pack' (*ibid.*, pp. 28–9). 'While I am looking at the same external phenomenon I can say with equal truth "this is one group of trees" and "these are five trees" or "here are four companies" and "here are 500 men." What changes here is neither the individual nor the whole or the aggregate, but my terminology. But that is only a sign that one concept has been replaced by another' (*ibid.*, p. 59). In order to count we need a unit for counting and that is provided by the concept. A concept 'is the unit with respect to which a number belongs to it. . . . The concept to which a number is ascribed in general isolates that which falls under it in a definite way. The concept "letters in the word one" isolates the *o* from the *n*, the *n* from the *e*, etc. The concept "syllables in the word one" picks out the word as a whole and as indivisible in the sense that its parts no longer fall under the concept' (*ibid.*, p. 66).

Counting is a case of establishing a one-to-one correspondence and that notion was defined in purely logical terms in the *Begriffsschrift*. Once we know that in order to count we require concepts as units we can see that counting establishes a correspondence between concepts. With these considerations we are well on the road towards a definition of the natural numbers, but they do not yet in themselves provide that definition. They merely state conditions any such definition would have to satisfy. When we say that the objects falling under a concept F can be correlated one-to-one to the objects falling under a concept G, then we are saying that the two concepts are *equinumerous*. But with such assertions we have not yet got to the numbers. For the numbers are objects, and here we have only statements about concepts. 'We have already settled that number words are to be understood as standing for independent objects. And that is enough to give us a class of propositions which must have a sense, namely those which express our recognition of a number as the same again' (F, p. 73). Frege tries to explain how this can be done by drawing on an analogous case in geometry. He writes:

The judgment 'line *a* is parallel to line *b*' or, using symbols,

$$a // b$$

can be taken as an identity. When we do this we obtain the concept of direction and say: 'The direction of line *a* is identical with the direction of line *b*.' Thus we replace the symbol // by the more generic symbol =, through removing what is specific in the content of the former and dividing it between *a* and *b*. We carve up the content in a way different from the original way and this yields us a new concept (*ibid.*, pp. 74–5).

Formally, it seems that the notion of direction can be defined in terms of the notion of parallelism; just as parallelism can be defined in terms of the notion of sameness of direction. However, this would be 'to reverse the true order of things. For surely everything geometrical must be given originally by intuition. . . . The concept of direction is only discovered at all as a result of a process of intellectual activity which takes its start from the intuition. On the other hand we do have an idea of parallel straight lines' (*ibid.*, p. 75). In logic there is no question of anything being given in intuition, but we can still speak of logical priority. And that determines that the notion of equinumerosity is prior to that of number.

The definition of the notion of direction in terms of the notion of parallelism seems initially appealing. If we consider it satisfactory, it gives

IN SEARCH OF LOGICAL OBJECTS

us a procedure that can also be applied to the definition of number. Starting with the notion of two concepts being equinumerous we can consider this equivalence:

(1) The concepts F and G are equinumerous if and only if the number of F's is the same as the number of G's.

Appealing as this idea may be, however, it leads to difficulties which can be brought out by considering the equivalent geometrical case a little more closely.

There are two difficulties with the attempted definition of direction in terms of the equivalence.

(2) The lines a and b are parallel if and only if the direction of a is the same as the direction of b.

The first is this. If we consider (2) a definition of the notion of direction, then we treat the sentence occurring before the words 'if and only if' as explaining the meaning of the sentence that follows it. But presumably we already connect some meaning with the words 'the same as,' i.e., the notion of identity. The question therefore arises: 'Are we not liable, through using such methods, to become involved in conflict with the well-known laws of identity?' (F, p. 76). Such doubts are based on a principle that Frege states explicitly in *Grundgesetze*: 'One must not explain a sign or word by explaining an expression in which it occurs while the other parts are known' (GG, vol. 2, p. 79).

Given that identity is explained by Leibniz's law of the substitutability (*salva veritate*) of identicals, we must ask whether (2) defines the notion 'the direction of a,' or '$D(a)$' for short, in such a way that it is substitutable everywhere *salva veritate* by '$D(b)$.' 'This task is made simpler by the fact that we are being taken initially to know nothing that can be asserted about the direction of a line except the one thing, that it coincides with the direction of some other line. We should have to show only that substitution was possible in an identity of this type' (F, p. 77). Frege believes that in fact this can be done without serious difficulty.

The second problem with the proposed definitions (1) and (2) turns out to be more serious. Frege argues that they fail to specify truth-conditions for identities such as

(3) the direction of line $a = q$

and

(4) the number of $F = q$

where q is neither a direction nor a number. An object is well defined only when we can determine whether any other object is identical with it or not. Propositions (3) and (4) should of course turn out to be false, but that conclusion cannot be derived from the proposed definitions (1) and (2). 'What we lack is the concept of direction; for if we had that we could lay it down that, if q is not a direction, our proposition is to be denied' (F, p. 78). Unfortunately, proposition (2) does not furnish us with the general notion of direction and analogously (1) does not furnish us with the necessary concept of number.

Frege concludes that the attempt to give contextual definitions of the notions of direction and number has failed. The question how they are to be defined must therefore be asked all over again. At this moment Frege makes a move that is not in any way prepared for by any of his earlier considerations: he introduces the notion of the extension of a concept.

> Seeing that we cannot by these methods obtain any concept of direction with sharp limits to its application . . . let us try another way. If line a is parallel to line b, then the extension of the concept 'line parallel to line a' is identical with the extension of the concept 'line parallel to line b'; and conversely, if the extensions of the two concepts just named are identical, then a is parallel to b. Let us therefore try the following type of definition: the direction of line a is the extension of the concept 'parallel to line a' (*ibid.*, p. 79).

For the very same reasons the attempt to define numbers in terms of equivalence (1) must also be rejected, an explicit definition of number is unavoidable and it can be given in analogy to that of the notion of direction:

(5) The number of a concept F = the extension of the concept 'being equinumerous to F.'

The conclusion Frege has reached provides him with the definition of number that he had set as his target for the *Foundations of Arithmetic*. With that definition he can define particular numbers starting with zero, the notion of the successor of a number, and the general notion of natural number. We need not concern ourselves with the details of his exposition.[21]

But there is something surprising and disturbing about the definition of numbers in terms of extensions of concepts in the general context of Frege's thought. He had originally reasoned that numbers as logical objects had to be defined contextually. It was presumably for this reason that he titled the relevant section of the book: 'To obtain the concept of Number, we must fix the sense of a numerical identity' (F, p. 73). But the conclusion of that section was that the attempted definition which fulfilled that condition could not legitimately be adopted. A definition of numbers as objects had been achieved, but the question remained whether that definition revealed numbers to be logical objects. The notion of the extension of a concept which is used in the definition appears out of nowhere on page 79 in the *Foundations of Arithmetic*. How is it to be understood? Is it a logical notion and why? In the *Foundations* Frege writes somewhat lamely: 'I assume that it is known what the extension of a concept is' (*ibid.*, p. 80). And he suggests that after all the notion might not be essential to the construction: 'I attach no decisive significance even to bringing in the extensions of concepts at all' (*ibid.*, p. 117; cf. also p. 80).

The issue is fully resolved only in the essay 'Function and Concept' of 1891 where extensions of concepts are characterized as special cases of value-ranges of functions and an argument is provided for considering value-ranges as logical objects. As far as the *Foundations of Arithmetic* is concerned, the book did not settle the issue it had tried to settle. It had begun with the problem of the definition of number. And even though that problem had been solved, the book had ended in a dilemma. It seems possible that Frege had envisaged the work as providing the desired logical analysis of numbers announced in the *Begriffsschrift*. But the book he wrote was at most a stepping-stone to that analysis. It clarified the issues, but did not resolve them.

V

The Analysis of Meaning

1 The development of Frege's views in the 1890s

The years that separate the *Begriffsschrift* from the *Foundations of Arithmetic* were a time of creative silence for Frege. Publishing almost nothing during those years, he rethought and deepened his philosophical and mathematical conceptions. With the completion of the *Foundations* another such period began. The silence was finally broken after six years with a string of five essays ('On the Principle of Inertia,' 'Function and Concept,' a review of Cantor, 'On Sense and Reference,' and 'On Concept and Object'). These were followed almost immediately by the first volume of the *Grundgesetze*, Frege's largest and most technical work. In its preface he writes:

> With this book I carry out an enterprise which I had in mind in my *Begriffsschrift* of 1879 and which I announced in my *Foundations of Arithmetic* of 1884. . . . The reason the execution has followed so late after the announcement lies in part in internal transformations of my logic which forced me to abandon a draft that was almost completed in manuscript (GG, vol. 1, pp. viii–ix).

He points out three major changes in his logic. First, he has added value-ranges to the system. 'The introduction of value-ranges of functions has been a major progress,' he says (*ibid.*, p. ix). The *Foundations of Arithmetic* had left unresolved the question why extensions of concepts should be considered logical objects. From 1891 onwards Frege treats extensions of concepts as special cases of value-ranges of functions. And he hopes to show that value-ranges are logical objects and thus to close the gap left in the argument of the *Foundations*.

In order to show why statements about value-ranges may be considered logical truths Frege sees himself forced to draw a distinction between two aspects of meaning, and this distinction constitutes for him the second

THE ANALYSIS OF MEANING

major change in his logic. Previously he had talked about the sign and what it signifies or, in Lotze's terminology, about form and content. In the essay 'Function and Concept' a new schema is proposed. In addition to the sign and what it signifies, a third element is recognized. What was formerly called the content of the sign is now called '*Bedeutung*' and the new third component is called the '*Sinn*' of the expression (commonly translated today as 'reference' and 'sense'). The distinction turns out to be vital not only for the proof that value-ranges are logical objects, but useful also in another respect: it allows one to separate the thought from its truth-value. In the *Grundgesetze* Frege writes as follows: 'Only intensive occupation with this book can teach how much simpler and more precise everything has become through the introduction of truth-values' (*ibid.*, p. x).

The analyses of the *Foundations* had turned on a distinction between objects and concepts, but had failed to determine the nature of the distinction. In 1891 Frege takes the issue up in the wider context of the distinction of objects and functions. The notion of function was originally introduced in the *Begriffsschrift* but the task of clarifying it had remained uncompleted. The third major innovation to which Frege refers in the preface of the *Grundgesetze* is therefore that 'the nature of functions in contrast to objects has been characterized more sharply' (*ibid.*).

These changes are introduced and explained in the five essays that precede the first volume of the *Grundgesetze*. In them mathematical, logical, and semantic considerations are closely intertwined. They reveal how philosophical concern with language can grow out of investigations into the foundations of mathematics and logic. The bonding together of these three areas of inquiry has remained characteristic of much of analytic philosophy and has very largely determined the kind of interest analytic philosophers have taken in language. For Frege the essays provide the background against which the systematic deduction of arithmetic from logic is to be carried out. With that background fixed, Frege thought he could bring to an end the mathematical and philosophical line of reasoning he had pursued for almost twenty years. The foundations of arithmetic would be secured once and for all, the logical, *a priori* character of arithmetic would be established beyond any possible doubt, and the falsity of empiricism would be exposed. To understand the essays that precede the *Grundgesetze* is therefore to understand Frege at the moment of his hoped-for triumph.

2 The relation between a theory and its constitutive concepts

The first of the essays of 1891 and 1892, 'On the Principle of Inertia,' is probably the least read even though it might be considered the most fundamental. The reason for the neglect is presumably that Frege's review of Ludwig Lange's book on *The Historical Development of the Concept of Motion and its Foreseeable End Result* seems at first sight to be concerned with issues that are far removed from the other logical issues of the period. It is true that Frege had a genuine interest in problems of physics, but the essay 'On the Principle of Inertia' is not entirely or even predominantly motivated by that interest. Its real focus is not the concept of inertia, but the nature of the relation of a theory to its constitutive concepts. And that issue is central to Frege's logical problems in the early 1890s.

In the *Logische Untersuchungen* Trendelenburg had written about theories and their concepts:[1]

> In the sciences every concept is preceded by a judgment or a series of judgments in which it has its justification and inner order. . . . The whole judgment of the Copernican world system precedes concepts like that of the path of the earth or the maximum and minimum distance of the sun.

He explicitly referred to Gruppe's views in this context. It is possible that Frege knew these remarks. In any case, in his essay 'On the Principle of Inertia' he defends a position that is identical with Trendelenburg's.

Lange's book, which occasions the essay, is concerned with a critique of Newton's justification of the principle of inertia and proposes an alternative to it. Newton's argument involves the notions of absolute motion, absolute space, and absolute time, which Lange considers objectionable. They make Newton's argument dependent on the metaphysical assumption of a transcendentally real space and time. Lange's own endeavor is to establish a justification of the principle of inertia in which the reference system of motion is merely ideal. Given a triple of points which proceed simultaneously and on their own from a given point, a co-ordinate system relative to which the paths of these points are straight lines is to count as an inertial system. It is Lange's claim that the principle of inertia can be explained in terms of this notion of an inertial system.

In response to Lange's account Frege remarks that the difference between the two attempted explanations is not as large as Lange considers it to be. In Newton's assumption of absolute space more is indeed contained than is necessary for the explanation of phenomena, but that does not make the notion completely transcendental. The contrary impression is due only to the shortcomings of Newton's exposition. He was under the spell of an empiricist epistemology. 'The flaws of his explanation are accounted for by his aversion to advancing anything as a hypothesis. He wanted to derive every step immediately from experience or from highest principles that were self-evident to him. And this procedure drove him into isolating that which allows comparison with experience only as a whole' (KS, p. 115). The term 'absolute space' does not gain its meaning from experience, but rather by its role in the theory, and it is only the theory as a whole that can be compared with experience. If we understand this point correctly, Newton's concept of absolute space does not seem so metaphysical.

When Lange criticizes Newton he is really guilty of the same methodological mistake as Newton. His criticism overshoots its target because of 'the separate consideration of the hypotheses which have meaning [*Bedeutung*] only as a whole' (*ibid.*). Lange's alternative exposition of the principle of inertia claims to be non-metaphysical because it is based on the supposedly experiential notion of points proceeding simultaneously and on their own from a point of origin. But the claim that something is a point moving on its own goes beyond experience just as much as the claim that a point is in absolute motion. Lange's attempt to give experiential foundations to the principle of inertia fails for the same reasons as Newton's. 'The flaw derives in both cases from the isolation of the hypotheses. It is only the totality of the principles of dynamics which as a hypothesis can be compared with experience' (*ibid.*, p. 116).

In order to gain a better understanding of the relations between a theory and its constitutive concepts Frege thinks that we must first acknowledge that every theory involves the construction of an appropriate language. That is true not only in physics but in all sciences, including logic (*ibid.*, p. 122). Any such construction must follow certain rules: 'In every science the decisive idea in the formation of the technical language is that laws must be expressed in the simplest, most complete, and most precise manner' (*ibid.*). He quotes Leibniz's statement that 'we ascribe motion to bodies according to those hypotheses through which the phenomena can be explained most adequately' (*ibid.*, p. 118). Leibniz should have said 'convention' or, better still, 'definition' instead of 'hypothesis,' Frege comments, since 'Conventions are really neither true nor false, but practical or impractical. One will always prefer that manner of speech in which the natural laws can be expressed most simply' (*ibid.*).

Every theory is cast in a language that necessarily contains conventional features. The terms in which our theory is cast are chosen because they allow the most adequate and simplest formulation of the theory. They are meaningful not because we can give them experiential content outside of the theory, but because they play an indispensable part in it.

According to Frege, Lange's important and correct insight is that the concepts of a theory are not given prior to and independently of the theory. He says that 'the elementary concepts are not given at the outset of scientific reflection,' but he takes this insight only as historical; he believes that the elementary concepts of science undergo a process of evolution. Frege, in contrast, holds that 'for the logical concept there is no development, no history' (*ibid.*, p. 122). The history of science consists in a series of attempts to grasp the correct theory and with it its constitutive concepts. 'The concept is something objective which we do not form and which does not form itself in us, but we seek to grasp it and, in the end, with luck we do so' (*ibid.*).

The correct part of Lange's claim that the elementary concepts are not given at the outset is for Frege the idea that 'as I would express it, they must be discovered through the work of logical analysis. . . . What comes first logically and systematically is not what comes first psychologically and historically' (*ibid.*, p. 124). Concepts are not given prior to the theories in which they function. To ignore the logical dependence of the concept on the theory is to fall into confusion. The quarrel over the reality of motion is often nothing more than playing with words (*ibid.*, p. 121).

> In the same sense in which one calls the stability of a length real (for instance, that of the standard meter at an unchanged temperature) we also consider the differences between different kinds of motion real. In both cases we have arbitrary conventions which are so tightly connected with the laws of nature that they are distinguished from other logically and mathematically possible conventions. If one wants to express this connection with the lawfulness of events through the word '*wirklich*' one must do so in both cases. Perhaps, the word '*objektiv*' is more suitable (*ibid.*, pp. 121–2).

The *Begriffsschrift* logic had been constructed on the assumption of the priority of judgments over concepts. The analysis of arithmetical statements in the *Foundations of Arithmetic* had proceeded on the principle that one should ask for the meaning of words only in the propositional context and not in isolation. The significance of the essay 'On the Principle of Inertia' is that it restates these doctrines in the form of the claim of the priority of a theory over its constitutive concepts. It is in the light of this consideration that the distinctions of function and value-range, concept and object, thought and truth-value, and sense and reference must be understood. Their justification and meaning lie not in the intuitive content they may carry, but in their indispensability for an adequate logical theory. The essays in which Frege elaborates these distinctions must be understood as engaged in the search for what is logically and systematically prior, according to a method that is outlined in the essay 'On the Principle of Inertia.'[2]

This point needs emphasis because Michael Dummett among others has claimed that Frege abandoned the principle that words have meaning only in propositional contexts around 1890 and that the new doctrines of the 1890s are irreconcilable with that principle. Dummett writes:[3]

> The apprehension of the central role of sentences for the theory of meaning was one of Frege's most fruitful insights. . . . Nevertheless it was an insight which Frege let slip, one which cannot consistently be reconciled with the views which he later held.

THE ANALYSIS OF MEANING

Since the doctrines of 1879 and 1884 are closely tied to Kantian presuppositions, such a change in Frege's views would signal a deep shift in his philosophical thinking. The essay 'On the Principle of Inertia' seems to rule out the possibility of such a shift around 1891. This impression is confirmed by Frege's methodological approach to the analysis of real numbers in 1903 and by his remarks to Darmstaedter in 1919. Both the later theory and the later practice seem to be fully in accord with the earlier assumptions.

It remains to be considered why the contrary impression might have arisen. Two reasons are usually adduced in support of this interpretation. One is that in the essay 'Function and Concept' Frege's explanation of concepts as functions and, as a consequence, of sentences as names of truth-values is incompatible with the claim that sentences play a central role in the theory of meaning. The other reason is that the doctrine of sense and reference is primarily a theory of referring expressions (proper names and definite descriptions) and so the problem of sentence meaning is subsidiary to it.

But if Frege's understanding of the relation of judgments and concepts did not change around 1891, then his account of concepts and functions and of sense and reference must be shown to fit into it. And the apparent plausibility of the contrary interpretation must be explained. This can be done if we take seriously the essay 'On the Principle of Inertia' as an indication of Frege's semantic and logical views in the 1890s.[4]

3 The assignment of logical structure

It may help first to consider the general issues the essay raises, before we try to determine how they bear on the distinction of function and object and the theory of sense and reference. In the remarks for Darmstaedter Frege says that the sentence is primary in his logic. The fact that sentences express thoughts and that these thoughts can be true or false is to be considered logically fundamental. The parts of the thought are reached only by analysis.

A sentence which expresses a thought will usually have a grammatical structure. The question is whether any logical significance should be attached to that structure. In the *Grundgesetze* Frege writes:

> Any sign, any word can be considered as consisting of parts; but we deny its simplicity only if the reference of the whole would follow from the reference of the parts according to the general rules of grammar or signification and if these parts also occur in other combinations and are treated as independent signs with their own reference (GG, vol. 2, p. 79).

In another place he adds: 'It is completely mistaken to believe that one

THE ANALYSIS OF MEANING

could never distinguish too finely. It is only detrimental to stress differences where they are irrelevant' (NS, p. 154).

These considerations seem to be in accord with the procedure of the *Begriffsschrift*, where Frege insists that the subject–predicate distinction is logically irrelevant and that in a logically perfect language only as much structure should be ascribed to sentences as is necessary to account for the logical relations between them. This suggests that we should distinguish between the grammatical structure and the logical structure of a sentence and between the words of the sentence and what we might call the logical constituents of the sentence which make up its logical structure. A sentence in ordinary language is composed out of words and has a specific grammatical structure, but these do not necessarily indicate the logically significant features of the sentence. In order to determine the logical constituents of a sentence and the logical structure of a sentence we must determine the place of the sentence in the logical network. One way of putting the point is to say that logical structure is not an absolute property of a sentence, but a relational one involving a sentence and a set of sentences relative to which structure is assigned. We need to assign to a sentence only enough structure to account for the logical relations between it and the other sentences in the set.

We may illustrate the point by a number of simplified examples:

(1) Given the sentence '$F(a)$' and the set of sentences

$$A = [F(a), G(b), R(m,n)]$$

no logical structure need be assigned to the sentence. It can be considered a logically simple name of a truth-value.

(2) Given the sentence '$-(F(a) \rightarrow G(b))$' and the set of sentences

$$B = [-(F(a) \rightarrow G(b)), F(a), R(m,n) \vee F(a)]$$

a logical structure must be assigned to the sentence such that '$F(a)$' and '$G(b)$' appear as logical constituents. They in turn can once again be considered as simple names of truth-values.

(3) Given the sentence '$F(a)$' and the set of sentences

$$C = [F(a), F(b), \exists x F(x), \forall y F(y)]$$

we must assign to '$F(a)$' a more complex logical structure. We describe the sentence as consisting of a name 'a' and a first-level predicate '$F(x)$.' In this analysis 'a' and '$F(x)$' are considered logically simple.

(4) Given the sentence '$F(a)$' and the set of sentences

$$D = [F(a), F(b), \exists \phi \phi(a)]$$

we must again ascribe a more complex logical structure to '$F(a)$.'

We describe the sentence as consisting of a first-level predicate '$F(x)$' and a second-level predicate '$\phi(a)$.' Those two expressions are considered logically simple.

It appears then that the amount of structure we assign to a sentence depends on the context in which this sentence is considered. In some contexts — such as (1) and (2) — we assign less structure to one and the same sentence than in others — such as (3) and (4). Moreover, different and even incompatible structures may be assigned to the same sentence — as in (3) and (4).[5]

In accordance with these principles Frege writes:

> Language has the means of letting now this, now that, part of the thought appear as subject. One of the best known is the distinction of active and passive forms. It is therefore not impossible that the same thought appears in *one* analysis as singular, in another as existential, and in a third as universal (KS, p. 173).

On the one hand, he allows that the thoughts expressed by sentences must themselves have parts and that, by combining words which express these parts, we can form a multitude of sentences expressing a multitude of different thoughts. 'This would be impossible if we could not distinguish parts in the thought to which sentence parts correspond, so that the construction of a sentence can be considered a picture of the construction of the thought' (KS, p. 378; similarly NS, p. 243). On the other hand, talk of part and whole with respect to the thought is said to be no more than a fitting metaphor (KS, p. 378). The signs can never fully express the character of the thought. 'This divergence of the expressing sign from the thought expressed is an inevitable consequence of the difference between that which appears in space and time and the world of the thought' (*ibid.*, p. 381). Thoughts are not spatio-temporal and hence we cannot talk of their parts in the same way as we can talk about the parts of spatio-temporal sentences. That is why even in a logically perfect language the same thought will have different sentences expressing it. It will then appear as if the same thought were composed of different constituents or composed of the same constituents in different ways. A sentence and its double negation, for instance, express the same thought (*ibid.*, p. 368). So do the sentences 'A and B' and 'B and A' (*ibid.*, p. 381). The sentences 'A,' 'A and A,' and 'A or A' all express the same thought (*ibid.*, p. 392).

If we consider thoughts as built up out of absolutely simple constituents, we will be forced to draw a distinction between elementary thoughts composed of constituents corresponding to simple names and simple predicates and thoughts which are the negations of such elementary thoughts. Atomism is always committed to such a distinction of affirmative and negative propositions, whether or not we can actually say which propositions are to count as the one or the other. Frege completely rejects an atomistic view of thoughts:

> One speaks of affirmative and negative judgments. Even Kant does that. Translating into my terminology one would distinguish between affirmative and negative thoughts: a distinction which is altogether unnecessary at least as far as logic is concerned. I do not know a single logical law in whose formulation it would be necessary or even advantageous to use this distinction. In every science in which one can speak of lawfulness one must always ask: which technical terms are necessary or at least useful for the precise expression of the laws of that science. What does not withstand this examination is of evil (KS, p. 369).

The passage shows that thirty years after writing 'On the Principle of Inertia' the methodological considerations of that essay are very much alive in Frege's mind. It is those considerations that block an atomistic conception of judgment for him.

4 Concepts as functions

The first use to which the doctrines of the essay 'On the Principle of Inertia' are put is in the clarification of the distinction between objects and concepts. From 1880 onwards Frege thought of numbers as individual things or objects. That was because we can assert identity of them; we can say that a number specified in such and such a way is just the same as a number specified in another way. As objects the numbers must be sharply distinguished from concepts. In the *Foundations of Arithmetic* Frege had pointed out the similarity between quantificational statements and number statements. To say that Venus has no moons seems to amount to the same thing as to say that the number of moons of Venus is zero. To say that the Earth has a moon and that it has no more than one moon seems to amount to saying that the number of moons of Earth is one. This might lead one to think of numbers as predicates of predicates just as Frege treats the 'there is' as a second-level predicate. He himself had toyed with the possibility but had rejected it because of his view that numbers are objects and not concepts.

But why should the distinction between objects and concepts be thought to be hard and fast? Why couldn't something play the role of object in one judgment and of concept in another? In the essay 'On Concept and Object' Frege struggles with this objection as raised by Benno Kerry. If there is a sharp distinction between the two, how is it to be explained and how can logical objects be constructed out of logical concepts?

Frege holds that the distinguishing mark of objects is that we can assert identity of them. It follows that we cannot make identity statements about concepts, if they are strictly distinct from objects. Unfortunately, that claim seems to conflict with the facts. Can we not assert that the concept 'horse' is different from the concept 'animal'? Frege's solution in the essay 'On Concept and Object' is to say that 'the concept horse' refers not to a concept, but to an object. That doctrine may not be incoherent, but it is highly paradoxical. Frege was willing to accept the paradox if it was necessary for preserving the separation of objects from concepts.

However, he realized that an explanation is required of why identity cannot be asserted of concepts. When he had first considered the distinction of objects and concepts, he said that a concept is 'nothing complete, but only the predicate of a judgment from which the subject is missing' (NS, p. 18). The characterization of concepts as predicates of possible judgments he had taken from the *Critique of Pure Reason* (B 94). But why did the Kantian conception imply that concepts are incomplete and that identity therefore could not be asserted of them? Frege's formulations seem also indebted to Lotze, who writes that the grammatical distinction of substantives, adjectives, and verbs represents a fundamental logical distinction:

> And it is hardly necessary to insist that the various characters impressed by language upon its material are the indispensable condition of the later operations of thought; it is obvious that neither the combination of marks into the concept, nor that of concepts into judgments of or judgments into the syllogism would be possible, if the matter of every idea were equally formless or were apprehended in the same form, if some of them were not substantival and did not express fixed and independent points of attachment for others which are adjectival, or again if others again were not verbal, exhibiting the fluid relations which serve to bring one thing into connection with another (*Logik*, pp. 12–13).

For Lotze 'the definite article . . . marks the word which it accompanies as the name of something to which we point' (*ibid.*, p. 11). But nothing can have a name made for it 'unless it has been thought of as identical with itself, as different from others, and as comparable with others' (*ibid.*, p. 19). For Lotze, then, as for Frege, it is substantives or names, expressions with definite articles, that stand for fixed, independent points of attachment about which we can make identity statements.

But Lotze's account of the difference between objects and concepts is merely metaphorical and not really explanatory. The same is true of what Frege says about the distinction much of the time. He even insists that one can only speak of it in metaphorical terms 'whereby one has to rely all the time on the good will and the understanding of the reader' (KS, pp. 269–70). The distinction between the complete and the incomplete is a 'fundamental logical phenomenon which must simply be taken for granted and cannot be reduced to anything simpler' (*ibid.*, p. 269).

Frege offers a variety of metaphors to make the difference between objects and concepts comprehensible. He writes, for instance: 'An object—e.g., the number two—can logically speaking not stick to another object—e.g., Julius Caesar—without some kind of cement which can itself not be an object' (*ibid.*, p. 270). At other times he uses the chemical metaphor that only an unsaturated solution can be saturated and that it must be saturated by something which is not itself an unsaturated solution. He also compares the analysis of a sentence with that of a line: a line cannot be split into two segments without a common point such that both segments have first and last points. Again, at other times, the distinction between object and concept is motivated by the traditional metaphysical distinction between substance and accident. The concept needs something to which it can adhere.

As he struggled with the distinction it must have been clear to Frege that it has a logical purpose which is not exhausted by these metaphorical descriptions. Even if we can talk about the distinction only in metaphorical terms, there must be non-metaphorical substance to it, if it is to be considered a fundamental fact of logic. I believe that Frege came to consider the idea that concepts are functions as the key to the issue. That idea forms the backbone of the essay 'Function and Concept'; it explains Frege's assumption that concepts are incomplete, that we cannot make identity statements about them, and that they must therefore be distinguished sharply from objects such as the numbers.

The logical notion of function had been introduced in the *Begriffsschrift*, but was not referred to at all in the *Foundations of Arithmetic*. Frege's failure to talk about the relation between concepts and functions in the latter work has given rise to the idea that the identification of concepts with functions of a particular kind is a belated afterthought of about 1891. But that is not correct. A function in the *Begriffsschrift* sense is not just a mathematical expression, but may also be a predicate or a relational expression (cf. BS, pp. 15–16). The later account differs from the *Begriffsschrift* in two respects. In the *Begriffsschrift* the distinction of function and argument is syntactic; it is drawn at the level of the sign. A function, Frege says, is part of a complex expression that is considered invariant relative to certain possible substitutions and an argument is the variable element in such substitutions. Also in the *Begriffsschrift* there is no sharp categorial distinction between functions and arguments. Although there is a suggestion that some expressions occur characteristically as functions, the distinction is actually made in a way that would allow every expression to play the role of function at one time and of argument at another time.

In the essay 'Function and Concept' the identification of functions as certain kinds of expression is rejected as a typical mathematical confusion

(KS, p. 126). A function is not a sign but what the sign signifies. What had formerly been called a function is now called a functional expression. Frege also now holds that there is a sharp distinction between names and functional expressions; the former stand for objects and the latter for functions. Objects are complete, functions are incomplete. One of the most notable features of the argument is Frege's readiness to proceed immediately from names and functional expressions to what those expressions stand for. Why should we assume that they stand for anything? No argument is provided. The step is presumably justified by the methodological considerations of the essay 'On the Principle of Inertia.' If the distinction between complete and incomplete expressions is fruitful and necessary for logic, it must represent something real or objective. Ontological worries about objects and functions are mere playing with words.

The claim that concepts are functions does not itself explain why concepts are fundamentally distinct from objects, why they are incomplete, or why identity cannot be asserted of them. It merely puts these problems in a larger context in which they eventually can be resolved. We must first ask what is the substance of the distinction between names and functional expressions? Behind that distinction lies the observation that in ordinary language not every combination of words forms a complex expression. Sentences and complex phrases have a definite grammatical structure. The substance of the claim that two objects cannot adhere together is that in general two names conjoined together do not form a sentence, but only a list. In most natural languages the grammatical structure of the sentence is determined by form-indicating words like articles and connectives or by the inflections of verbs and the cases of nouns and pronouns. Frege holds that we can divide the words of our language into those which determine the structure of complex phrases in which they occur and those which do not. The latter are Frege's names, the former his functional expressions. An incomplete expression is for him a structure-determining expression, whereas a complete expression is not.

It is presumably not sufficient for him to show that a distinction between complete and incomplete expressions can be drawn in some natural languages. That would not establish the logical necessity of the distinction. He must show that every language presupposes a distinction between complete and incomplete. Assume that in our language the proposition that 'five is a prime' is expressed through the juxtaposition of the two complete expressions 'five' and 'the concept prime.' Where is the incomplete element in this sentence? Frege argues that the incomplete element is still present, but hidden. For when we say 'five—the concept prime' we mean to say that five *falls under* the concept prime and the relation of falling under is represented by the juxtaposition of the two complete expressions. It is the juxtaposition of the signs which here incorporates the incompleteness or unsaturatedness. Frege concludes that 'we can easily recognize that the difficulty which lies in the unsaturatedness of a thought constituent can be shifted, but it cannot be avoided' (KS, p. 178). And in so far as it cannot be avoided, it represents a logically necessary fact. At first, the distinction between the complete and the incomplete can be considered as merely conventional or pragmatic, but the need for the distinction justifies us in concluding that it stands for something real or objective.

A mathematical function is usually represented by an expression containing variables that indicate places into which number terms may be substituted. If we make appropriate substitutions the whole expression represents a number; it represents the number which is the value of the function for the chosen numbers as arguments.

> A functional expression always carries with it empty places for the argument (at least one such place), which in analysis is usually indicated by the letter 'x' which fills the empty places. But the argument must not be considered part of the function and the letter 'x' is therefore also not part of the functional expression. One can therefore always talk of an empty place in the functional expressions in so far as that which fills it does not really belong to the expression. Accordingly I have called the function unsaturated or in need of completion because its name must be supplemented with a sign for an argument in order to obtain a completed reference (NS, p. 129).

Originally Frege had called the distinguishing characteristic of concepts their incompleteness; later he often spoke of their unsaturatedness. The change in terminology is connected with his attempt to explain the phenomenon of incompleteness. Concepts are incomplete because they are functions and functions are unsaturated in the sense described in the quotation above.

The nature of a function $f(x)$ 'shows itself in the connection which it establishes between the numbers whose expressions are substituted for "x" and those numbers which then constitute the reference of the expression' (GG, vol. 1, p. 5). We can also say that functions correlate arguments with values (KS, p. 277). In other words, functions are not inert entities: they connect and correlate. They 'perform functions' in the contexts in which they occur. Functional expressions do not possess completed references, but we are still entitled to say that they have reference in so far as they are necessary constituents of our language.

There is here clearly an important difference between Frege's understanding of concepts and that of the philosophical tradition. The traditional view has it that concepts are non-linguistic items which can be apprehended and identified in various ways and which can therefore also be referred to in different ways. Thus, we might think that redness is a concept which can be picked out either as the reference of the predicate 'is red,' or through the general term 'redness,' or by means of the expression 'the concept red.' Kerry's objection to the Fregean doctrine rests on this conception. Frege's reply is that concepts can be picked out only as the references of expressions that play a certain role in our language. It is once again the views of the essay 'On the Principle of Inertia' that explain this claim. He writes: 'Kerry thinks that one cannot found logical conclusions on linguistic distinctions; but in the way I do this, no one who draws such conclusions can avoid it since we cannot communicate without language' (KS, p. 169).

This doctrine has surprising and paradoxical consequences. Since the essence of a function or a concept is not separable from the role it plays, we cannot name functions or concepts. Not only can we not pick out a function with a name, but we also cannot give it a name once we have picked it out. Otherwise we might first point to the part of the sentence 'five is a prime' which is not identical with 'five' and then name the concept referred to by that part with a new name, such as 'Herbert Hoover' or 'the concept prime'; in that case functions could still be said to have names. However, this procedure would not be successful naming, because the expression 'Herbert Hoover' would not be able to play the same role in sentences in which it occurs as our original concept expression plays; it would therefore not name a concept. We may not be able to do without words like 'the concept,' but we must always be aware of their inappropriateness (NS, p. 130). Even the expression 'the reference of the concept word A' must be rejected, since the definite article before 'reference' points to an object and denies the predicative nature of the concept (*ibid.*, p. 133).

> To a part of a thought in need of completion . . . there also corresponds something in the realm of reference. But it is wrong to call it concept, relation, or function despite the fact that we can hardly avoid doing so. . . . When we use the words 'concept,' 'relation,' 'function' . . . we miss what we are really aiming at (*ibid.*, p. 275).

The predicament in which language here finds itself surfaces in the question what the expression 'the concept horse' does stand for if it cannot stand for a concept. In the essay 'On Concept and Object' Frege concludes that it must stand for an object, but he does not say what kind of object it is that the expression refers to. From a passing remark in a footnote to the *Foundations of Arithmetic* it appears that the object would not be the extension of the concept and that Frege at one moment contemplated the possibility of the numbers as such objects rather than as extensions of concepts. But by 1891 he has definitely abandoned that idea. The solution to the puzzle is to be found in a posthumously published fragment written between 1892 and 1895, i.e., shortly after the essay 'On Concept and

Object.' In that fragment Frege says, as quoted, that the expression 'the concept A' is illegitimate; he no longer holds it to refer to anything.[6]

While this removes an unnecessary difficulty from Frege's account it does nothing to alleviate its overall paradoxical character. Frege is quite conscious of that fact when he writes: 'Whenever I want to speak of a concept language imposes on me with almost inescapable force an unfitting expression through which the thought is obscured, one might almost say distorted' (NS, p. 130).

The problem is that we need to use expressions like 'the concept A' when we are criticizing ordinary discourse or when we are trying to set up a precise logical language. The nature of the dilemma can be brought out in this way: since a sharp distinction is to be made between objects and functions, we must also distinguish sharply between functions which take objects as arguments and those which take functions as arguments. This forces upon us a sharp distinction of levels of functions. First-level functions are functions of objects, second-level functions are functions of first-level functions. Third-level functions, in turn, are functions of second-level functions. By means of expressions for higher-level functions we can make statements about functions. 'But through this the difference between objects and functions is not at all obliterated' (KS, p. 173).

Let us now call any distinction between two items a categorial distinction when there exists no predicate that can *meaningfully* be predicated of each. Frege's system of distinctions between objects, first-level, second-level, and third-level functions is then clearly a system of categorial distinctions. It is a peculiarity of categorial distinctions that they cannot be stated in the theory in which they obtain. In order to state the difference between an object and a function, for instance, we would have to say that item x is an object but not a function and that item y is a function but not an object. But that statement itself presupposes that we have a concept of 'object' that is true of item x but false of y, and a concept of 'function' that is false of x but true of y. That means that both concepts would have to be meaningfully predicable of x and y; but by assumption there is no such concept.

The same conclusion can be reached for any other theory in which categorial distinctions are made. Russell's theory of types in the form in which it is developed in *Principia Mathematica* (as a semantic rather than a merely syntactic theory) is a case in point. Kurt Gödel was the first to notice that this theory cannot be formulated in accordance with its own principles.[7] The unstatability of categorial distinctions does not make such distinctions illegitimate. It merely poses a problem of how they can be explained and justified. The distinctions may be describable in another language or theory, such as the meta-theory of the theory in which the categorial distinctions occur. In that case a strong argument is required to show why the same difficulty does not recur in the new language or theory.

For instance, we might think that the Fregean distinction between objects and functions can be explained by talking about their respective linguistic expressions. But in Frege's account names are just as complete as the objects they stand for and functional expressions just as incomplete as the functions they stand for. The distinction between names and functional expressions is no less a categorial distinction than that between objects and functions.

It now becomes clear why Frege thought our attempts to describe the difference must always remain metaphorical. That means that for him semantic considerations can never constitute a rigorous science; they can at best serve as auxiliary devices for the critique of ordinary language or the construction of a new logical language. In the *Tractatus* Wittgenstein argues for similar reasons that the logical form of propositions cannot be stated but can only show itself (4.121), and that the attempt to make logical form explicit is only an elucidatory undertaking (6.54).[8]

5 Functions and value-ranges

If we follow Frege in regarding concepts as functions, we must determine what correlation between arguments and values is established by concepts. A mathematical function correlates numbers with numbers; a concept correlates the objects of which we predicate it with what? To settle that question is for Frege to settle the question what a concept term refers to.

> In order to explain this I call to mind a circumstance that seems to speak in favor of the extensional logicians and against the intensional logicians. It is that concept terms can be substituted for each other in every sentence without affecting its truth if the same extension of a concept corresponds to them. That means that concepts behave differently as far as inference relations and logical laws are concerned only in so far as their extensions are different (NS, p. 128).

The matter must be put in this roundabout way because the relation of identity is not thinkable of concepts (*ibid.*, pp. 130–1). 'But even if the relation of identity is thinkable only of objects, there is a similar relation between concepts' (*ibid.*, p. 131). Frege has in mind the second-level relation which obtains between two concepts when everything that falls under the first also falls under the second and vice versa. If that relation obtains we may say that the extensions of the two concepts are identical.

Frege's understanding of concepts is thus purely extensional. Logic is primarily concerned with truth. Truth is essential to it. 'It is of no concern to logic how one thought comes out of another thought irrespective of their truth-values' (*ibid.*, p. 133). It is for this reason that truth is the primary object of logic, that in logic we aim at truth, and that truth or falsity is that aspect of the meaning of a sentence which we call '*Bedeutung*' (commonly translated as 'reference' or 'denotation'). We can therefore say that a concept correlates objects with truth-values. That is the *Bedeutung* of a concept. The characterization of truth-values as the *Bedeutung* of sentences is thus inseparable from the characterization of concepts as functions.

Dummett has argued that the doctrine that truth-values are the *Bedeutungen* of sentences signals Frege's abandoning of the context principle. This development, he writes, was 'an almost unmitigated disaster. . . . The most disastrous effect of the new doctrine was the abandonment of one of Frege's most important insights, that of the central role of sentences in the theory of meaning.'[9] But how can this be correct, if Frege's claim that the *Bedeutung* of a sentence is the truth-value and that concepts are functions that correlate objects with truth-values is built on the very assumption that truth is fundamental to logic? That must mean that sentences are fundamental as linguistic means for referring to truth. The assumption that the sentence remains central in the account of meaning Frege develops in the 1890s is confirmed by the explanation he gives of concepts as truth-functions. Since concept expressions are incomplete, their meaning cannot be determined except by considering the role they play in complex phrases and sentences. The question what the reference of a concept expression is can be answered only if the reference of the sentence in which the concept expression occurs is already determined.

The fact that functions are not separable from their role is the major reason for introducing the value-ranges. Since we cannot speak of the identity of functions, number statements cannot be interpreted as statements about functions. And yet they seem to involve functions or concepts. The resolution of the dilemma is to postulate objects that correspond to functions.

> The method of analytic geometry supplies us with a means of intuitively representing the values of a function for different arguments. If we regard the argument as the numerical value of the abscissa, and the corresponding value of the function as the numerical value of the ordinate of a point, we obtain a set of points that presents itself to intuition (in ordinary cases) as a curve. Any point on the curve corresponds to an argument together with the associated value of the function (KS, p. 129).

The function itself is not an entity; its nature is to correlate arguments with values. But through the function such a correlation is established and that correlation can be considered an entity with its own criteria of identity. This entity is called a value-range. The value-range of a function is the equivalent in Frege's theory of what mathematicians call functions (cf. KS, p. 130). If we compare Frege's theory with standard mathematical writings we find that it is Frege's functions that lack an equivalent there

THE ANALYSIS OF MEANING

and not, as is widely supposed, his value-ranges. For Frege the functions of the mathematicians cannot be considered fundamental mathematical notions; in order to explain what such functions are we must introduce the more basic notion of Fregean functions as something incomplete or unsaturated.

If we consider concepts as functions, then concepts will also possess corresponding value-ranges. And they can be identified with what are generally called extensions of concepts. Frege holds that 'the fundamental logical relation is that of an object falling under the concept: all relations between concepts can be reduced to it' (NS, p. 128). Just as value-ranges are less basic than their corresponding functions, so extensions of concepts are less basic than their corresponding concepts. 'The extension of a concept does not consist of objects falling under the concept, in the way in which, e.g., a wood consists of trees; it attaches to the concept and that alone. The concept thus takes logical precedence over its extension' (KS, p. 210). In accord with what Frege says about the relation of his functions to what mathematicians call functions, we can now say that when philosophers talk about concepts as entities which possess identity conditions they are talking not about Fregean concepts but about their corresponding extensions. The Fregean doctrine is distinguished from the usual philosophical account by the claim that concepts in the philosophical sense are not fundamental and must be explained by means of a new and deeper notion, that of the Fregean concept. If we read Frege in this way, some of our initial puzzlement over the doctrine of the incompleteness of concepts seems to disappear.

6 *Value-ranges and set theory*

With the characterization of extensions of concepts as value-ranges Frege believes he has answered one of the questions left over from the *Foundations of Arithmetic*. All that now remains is to establish that value-ranges are *logical* objects.

The new view shows what extensions are and it shows that they are not what they are commonly believed to be. It is commonly held that extensions of concepts are classes or sets, but in Frege's eyes the notions of class and set employed in philosophical and mathematical discourse are highly confused. Sometimes they are used in a logical sense and they then correspond to what Frege understands as the extension of a concept; they are then value-ranges of truth-functions. At other times they are understood as composed of their elements. Thus, Cantor defines a set as 'a collection into a whole of definite, distinct objects of our intuition or of our thought.'[10] He identifies physical space with a set of points. He also says that sets are formed from objects by abstraction. And then again he writes: 'By a manifold or set I understand any multitude that can be thought of as One, that is, every class of definite elements such that this class can be tied into a whole by a law. I believe that I have hereby defined something which is related to the Platonic εἶδος or ἰδέα.'[11] It is clear from this last remark that Cantor intends his sets to be abstract, logical notions. But in Frege's eyes he does not fully succeed. In reviewing Cantor's 'Mitteilungen zur Theorie des Transfiniten' Frege says that Cantor 'is unclear about what is to be understood by "set," though a faint glimmer of the right conception shines through' (KS, p. 164).

Mathematicians' lack of a proper grasp of the notion of set is also castigated by Frege in other writings. In 1895 he gave the fullest expression of his objections in critical remarks on Schröder's recently published *Algebra der Logik*. His attention there is almost exclusively directed towards the notion of manifold that he considers basic to Schröder's construction. For Schröder a manifold is composed of its elements; but is such a concept really useful in logic? It is certainly not a logical notion in Frege's sense, since it is not derived from the notion of truth as the notion of value-range is. If we construct sets out of elements we must ask whether there can be an empty set, whether a set containing exactly one element can be distinguished from that element itself, and whether the relation of an element to a set is transitive. If sets are built up out of elements what is left when all the elements are removed? What distinguishes a heap of one brick from that brick itself? And if a brick is a manifold of grains of sand and a wall a manifold of bricks, is the wall not a manifold of grains of sand? For the purposes of the foundation of arithmetic we must have empty sets, unit sets distinct from their elements, and an intransitive relation of elementhood. Can these requirements be fulfilled by the intuitive notion on which Schröder and other mathematicians rely?

The situation is different when we turn to value-ranges and extensions of concepts in Frege's sense. An empty set corresponds to an extension in which all arguments are correlated with the truth-value 'false.' Such an extension is composed of as many particular correlations as the universal set or any other set. Extensions are not more or less filled. A unit set corresponds to an extension in which exactly one object is correlated with the truth-value true; such a correlation is clearly distinct from the one element of the unit set. The relation of an object to the extension of a concept to which it is said to belong is also clearly not transitive. Extensions of concepts fulfill the requirements of logic.

In his review of Schröder's *Algebra* Frege tries to show that the intuitive notion engenders difficulties which are wholly avoidable in logic. It leads Schröder to confuse the relation of an object belonging to an extension with that of one extension being part of another. He confuses elementhood and class inclusion and as a result is led to question the Boolean notion of a universe of discourse.

Schröder assumes an empty class o and he says of it that 'o is called a domain which stands to every domain *a* in the relation of inclusion, a domain which is contained in every domain of the manifold.'[12] And elsewhere he says: 'The "nothing" is even subject to every predicate; the nothing is black and it is also not black; for the null-class is contained in every class' (*ibid.*, p. 238).

Let us now imagine a manifold *a* which contains all classes. Then it seems to follow that there must also be a class *A* which contains only *a*. But because the null-class is contained in every domain, *A* must also contain o. From the definition, it would therefore follow that $a = 0$ (*ibid.*, pp. 245–6). Schröder concludes that the Boolean universe of discourse is too large a manifold and inadmissible. We must distinguish different manifolds. The primary manifold is a manifold of objects such that 'among its elements which are given as "individuals" there are no classes which contain elements of the same manifold as that of individuals' (*ibid.*, pp. 246ff). The next manifold consists of classes of elements of the primary manifold. In this way a whole hierarchy of manifolds is created.

Church has pointed out that Schröder's reasoning leads him to construct a simple theory of types.[13] While that is true, it is also true that Schröder's reasoning is confused. Frege's objections concentrate on these confusions. He argues that Schröder's claim that the null-class is contained in every class can only mean that the null-class is a subclass of every class, it does not mean that it is an element of every class. Schröder has been led into his difficulties because his intuitive conception of set confuses these two different relations. Frege concludes:

> One might easily get the impression from these deliberations that in the quarrel between extensional and intensional logicians I take the side of the latter. I do indeed hold that the concept logically precedes its extension and I consider as a failure any attempt to build the extension of the concept on something other than the concept, as a class built on the individual things. In this way one might reach a calculus of domains, but never a logic. Nevertheless I am closer to the author in certain respects than I am to those whom one might call in contrast intensional logicians (KS, p. 210).

Frege's warning is useful, since there is a tendency to think of concepts as intensional and of classes or extensions of concepts as extensional and to see that as the decisive difference between the two sorts of things. For Frege concepts and extensions of concepts are equally extensional; the difference between them is simply that concepts are incomplete whereas their extensions are not. That has the important consequence that statements about concepts can often be replaced by statements about their corresponding extensions without loss of content. The Fregean distinction of levels of functions seems at first sight to generate an infinite hierarchy corresponding to the infinite hierarchy of propositional functions in

Russell's theory of types. But in fact Frege employs only one third-level function in the whole system of the *Grundgesetze*. He comments:

> One might think that this is not nearly sufficient, but we will see that we can make do with this function and that it will only occur in a single sentence. It might be noted here that this frugality is made possible by the fact that functions of second level can be represented in a certain way by functions of first level whereby the functions that appear as arguments of the former are represented by their value-ranges (GG, vol. 1, p. 42).

7 *The logical justification of value-ranges*

According to what has been said so far, two functions $f(x)$ and $g(x)$ which have the same values for the same arguments will correspond to the same value-range. Frege explains in 'Function and Concept':

When we write:

$$x^2 - 4x = x(x - 4)$$

then we have not set one function equal to another, but we have only set the values of the functions equal. And when we understand this equation as being valid whatever is substituted as an argument for x, then we have expressed the generalization of an equation. But we can also express this by saying 'the value-range of the function $x(x - 4)$ is equal to that of the function $x^2 - 4x$' and thus we have an equation between value-ranges. That it is possible to regard the generalization of an equation between values of functions as an equation between value-ranges does not seem to me to be capable of proof, but must be considered as a basic law of logic (KS, p. 130).

What is here illustrated by means of a particular example is in fact nothing other than the basic law V of the *Grundgesetze*. If '$\dot{x} f(x)$' stands for the value-range of the function $f(x)$, we can write that law informally as

(1) $\forall x(fx = gx) \leftrightarrow (\dot{x} fx = \dot{x} gx)$

The question remains why Frege should consider this law a basic logical truth. It is not immediately clear how that claim could be argued. On the contrary, it might look more plausible to deny that (1) could possibly be a logical truth since the sentence on its left side makes a statement about functions and the sentence on the right side is about value-ranges. Frege insists on a sharp distinction between functions and objects, so it might be argued that the one side of the equivalence is made true by a quite different state of affairs from what makes the other side true. Even if (1) is true, there is then no reason to think that it is logically true.

THE ANALYSIS OF MEANING

It has not generally been noticed that in the essay 'Function and Concept' Frege offers an explicit justification for taking (1) to be logically true. There are two possible reasons why it has been overlooked: the argument is very concise and it is most unexpected. It deserves special attention not only because it shows how Frege tries to fill the lacunas left by the considerations of the *Foundations of Arithmetic*, but also because it shows how the theory of sense and reference is intertwined with Frege's philosophy of mathematics.

The connection between the two lies in the fact that in Frege's logic no distinction is made between identity and truth-functional equivalence of sentences. This doctrine is already contained in the *Begriffsschrift* (cf. BS, pp. 13ff, 54, 55). In the later writings it is explained by the assumption that the reference of a sentence is a truth-value. Material equivalence of sentences is therefore literally identity of their reference. According to this doctrine proposition (1) is strictly speaking an identity statement:

(2) $\forall x (fx = gx) = (\acute{x} fx = \acute{x} gx)$

The question whether (2) is logically true is therefore a special case of the question under what conditions an identity statement is logically true.

In the essay 'On Sense and Reference' the distinction between the two aspects of meaning is motivated at the outset by considering the question how identity statements can be true or false. One might consider this approach to the distinction of sense and reference an ingenious didactic device. My suggestion here is that Frege introduces his essay in this way because the problem of the truth of identity statements is basic to his philosophy of mathematics.

In order to understand this intertwining of the theory of meaning with the philosophy of mathematics one must first turn back to the *Begriffsschrift* account of identity. In the essay 'On Sense and Reference' Frege argues that this account is unsatisfactory. I interpret this to mean that the account is unsatisfactory first of all because it cannot explain why proposition (1) is logically true. In any case, the later account is constructed in contrast to the earlier one and it is to the *Begriffsschrift* account that we must turn first.

Frege writes in the *Begriffsschrift*:

> The same content can be determined in different ways; but that in a particular case two ways of determining it really yield the same result is the content of a judgment. Before this judgment can be made, two distinct names, corresponding to the two ways of determining the content, must be assigned to what these ways determine. The judgment, however, requires for its expression a sign of identity of content, a sign that connects these two names. From this it follows that the existence of two names is not always merely an irrelevant question of

form; rather that there are such names is the very heart of the matter if each is associated with a different mode of determining the content (BS, pp. 14–15).

What is called the content in this passage must correspond to what is called the reference of an expression in the later terminology. This content must be distinguished from the way in which such a content is determined. Frege assumes that the way the content is determined is reflected in the name associated with the content. Thus he argues that the same point in a geometrical construction can be named in two different ways, and that reflects the fact that the point is given in two ways, once intuitively and once as the point which satisfies a particular geometrical description. The first name is a simple letter, the second a definite description.

Frege maintains in the *Begriffsschrift* that an identity statement is a statement about names rather than their contents. In general signs are merely representatives for their contents, but in the identity statement they suddenly represent themselves. 'Thus the introduction of the sign for identity of content is connected with a certain ambiguity in the meaning of all signs in that they sometimes stand for their content, sometimes for themselves' (*ibid.*, p. 14).

The doctrine Frege develops here is directly derived from Lotze and indirectly from Leibniz. In his *Logik* Lotze presents his theory of identity in connection with the question of the status of arithmetic truths. Like Frege, he rejects the Kantian view of arithmetic as synthetic *a priori*. Kant had argued in that way because he considered the principle of identity to be the highest logical principle and clearly the laws of arithmetic could not be derived from that law alone since it 'contains not the slightest hint of a contrast between form and content' (*Logik*, p. 584). The principle says merely that everything is identical with itself; that is an 'infertile tautology' from which no substantial mathematical truth can be derived. 'I need not enlarge on the fact that it is this possibility of equating the different, not the barren application of the logical principle of identity, that is the motivating force of all fruitful mathematical reasoning' (*ibid.*).

This observation is for Lotze the reason for reflection on the notion of identity:

> Equations such as $\sqrt{4} = 2$ determine the specific quantitative value that results from applying a calculating operation to a given magnitude or, as in $\sqrt{a \cdot b} = \sqrt{a} \cdot \sqrt{b}$, they say that one reaches the same result when one carries out formally distinct operations in a given order or in combination. . . . In each case the value of the mathematical procedure does not lie solely in the discovered equality of the result, but in the fact that different ways have led to the same goal, that it was possible, so to say, to equate the different (*ibid.*, pp. 583–4).

THE ANALYSIS OF MEANING

In his 'Dialogue on the Connection Between Things and Words' Leibniz had made a similar point:[14]

> For even though characters are as such arbitrary, there is still in their application and connection something valid which is not arbitrary; namely, a relationship which exists between them and things, and, consequently, definite relations among all the different characters used to express the same thing.

The account of identity given by Leibniz, Lotze, and the Frege of the *Begriffsschrift* seems to come to the following: a distinction must be made between the thing named by an expression and the manner in which that expression names the thing. The way in which an expression names a thing is determined by the simpler terms out of which the expression is composed and their combination. The expression 'the evening star,' for instance, determines the thing it stands for as that star which is (first) visible in the evening. An identity statement is true because the thing named by the two expressions is the same; a true identity statement is informative because the two expressions determine the thing they name in two different ways. In order to explain the informativeness of identity statements we must take them to be about the way the expressions stand in relation to the things named and not just about the things themselves.

This account is rejected in Frege's writings after 1891 and is replaced by the explanation in terms of sense and reference. We can suggest several reasons for this change. The early account depends on the assumption of an ambiguity in the use of our signs. They are said to stand sometimes for the things named and sometimes for themselves. This doctrine has certain technical disadvantages. For instance, consider the sentence

(3) Object a has property F and a is identical with b; therefore b has the property F.

According to the *Begriffsschrift* the first occurrence of 'a' stands for the objects named, the second for the sign itself, whereas the first occurrence of 'b' stands for the sign itself and the second for the object named. Consider now a substitution of two other terms 'n' and 'm' for 'a' and 'b.' Are we to make the substitutions for both occurrences of 'a' and 'b' or for only one of them? Either decision leads to awkward difficulties. For this reason Frege later insisted on a rigorous distinction between the use and the mention of a sign. It can be said that drawing this distinction led Frege to abandon his earlier account of identity. On the other hand, it is not clear that the early account could not have been modified to accommodate a strict use–mention distinction.

The second reason for the change in Frege's thinking may have been the fact that he says in the *Begriffsschrift* that a true identity statement in which the same thing is named in two different ways is 'synthetic in the Kantian sense' (BS, p. 15). But in order to establish the analyticity of arithmetic Frege must clearly show that identity statements in which the same thing is named in two different ways can be analytic. Again, it is not evident that this requires a switch to the theory of sense and reference. Could one not argue that such identity statements might after all be analytic? That was Lotze's view.

Frege himself gives a different and even more inconclusive reason for the switch from the earlier theory. In the essay 'On Sense and Reference' he writes:

> What one wants to say by $a = b$ seems to be that the signs or names 'a' and 'b' designate the same thing and thus the signs themselves would be talked of; a relation between them would be asserted. But this relation between the names or signs would obtain only in so far as they name or designate something. It would be mediated through the connection between each of the signs and that which is designated. But this is arbitrary. No one can be forbidden from using any arbitrarily chosen producible event or object as a sign for something. Thus, the sentence $a = b$ would no longer concern the subject matter itself but only our mode of designation; we would express no real knowledge in it (KS, p. 143).

This argument is peculiar because it contains nothing that Frege had not been fully aware of at the time of the *Begriffsschrift*; it merely repudiates what the *Begriffsschrift* account asserts. And that earlier account actually appears more plausible. While it is true that we can arbitrarily choose simple names for things, it is not true that the designation of a complex expression is arbitrary in the same way. Once the designations of the simple terms occurring in it are determined, the designation of the complex expression is fixed by those designations and by the manner in which the simple terms are combined. The *Begriffsschrift* account, like that of Leibniz and Lotze, depends on this very insight.

In 'On Sense and Reference' Frege goes on to say that the identity statement can express knowledge only if the difference in the signs corresponds to 'a difference in the way in which the designation is given' (*ibid.*, pp. 143–4). This 'mode of determination' is contained in 'the sense of the sign' which must be distinguished from its reference.[13] The question is how the 'mode of determination' that Frege had spoken of in the *Begriffsschrift* is to be distinguished from the 'mode of determination' that he speaks of in 1891. The important difference between them seems to be that the former is a property of the signs whereas the latter is not necessarily reflected in the sign. In each account Frege assumes that objects are given to us in a certain way, that they present themselves to us under this or that aspect. In the *Begriffsschrift* the only question is how the way in which an object is given is reflected in the signs themselves. The identity statement is about the way in which the signs designate their objects. In the later account it is assumed that the way in which an object is given is not necessarily represented in the mode of designation of the signs. What makes an identity statement trivial or informative is not how the signs designate, but how their designation is determined. For that reason Frege allows that the name 'Aristotle' may have different senses for different speakers, since the object designated may be determined for them in different ways. The earlier account left no room for such a claim.

The distinction between the earlier and the later account is a distinction between a strictly linguistic explanation of identity statements and one that is given in cognitive terms. The Fregean notion of sense has its roots not in the theory of meaning, but in epistemology. It remains to be seen whether this conflation of semantics with epistemology was a happy one.

We can also interpret the shift as a change from a Lotze–Leibniz account to a Kantian one. According to Kant an analytic judgment is one in which the constitutive concepts are analysed and it is determined that the predicate concept is contained in the subject concept. In synthetic judgments 'I must have besides the concept of the subject something else (X), upon which the understanding may rely' (*Critique of Pure Reason*, A 8). We must therefore distinguish between the level of judgment and concept and the level of designated objects. In synthetic judgments 'we have to go outside these concepts, and call in the aid of intuition which corresponds to one of them' (*ibid.*, B 15). In the Kantian doctrine the concepts that we analyse or compare with objects may be explicitly represented in the formulation of the sentence expressing the judgment, but they need not be so expressed. 'If I say: "All bodies are extended," I have not amplified in the least my concept of body, but have only analysed it, as extension was really thought to belong to that concept before the judgment was made, though it was not expressed' (*Prolegomena*, 267). A predicate may be thought in the subject 'though not so clearly and with the same consciousness' (*ibid.*).

There has been much puzzlement about the origin of the Fregean theory of sense and reference. It has been noticed that in certain respects it is similar to the medieval theory of supposition and that it also shares common features with the distinction of connotation and denotation as elaborated by Mill. My suggestion here is that the root of Frege's distinction is to be found in neither of these sources, but that it lies in the Kantian distinction between concepts and objects.

My further suggestion is that the shift from the Lotzean to the Kantian conception, reflected in the shift from the *Begriffsschrift* to the later account of identity, is motivated by none of the reasons that have so far been offered, but by the requirements of Frege's philosophy of mathematics. The attempt to show that proposition (1) is a logical truth is what forced the shift from the semantic to the epistemological level that is to be found in the development of Frege's thought.

In order to grasp the need for this change one must turn back to the point at which the argument for the analyticity of arithmetic had been left in the *Foundations*. Frege had approached the problem of the logical definition of numbers by considering the analogous problem of the definition of the notion of direction. He had suggested that that notion might be introduced through the equivalence

(4) $a // b \leftrightarrow (D(a) = D(b))$

and that the definition of numbers could be achieved by the structurally similar equivalence

(5) Fx and Gx are equinumerous $\leftrightarrow (N_xFx = N_xGx)$

He had initially argued that logical objects must be introduced by equivalences of this kind since they had to be given both contextually and by laying down identity conditions for them. If (4) and (5) were to count as definitions they were certainly definitions of an unusual kind. 'Admittedly, this seems to be a very odd kind of definition to which logicians have not yet paid enough attention' (F, p. 74). Their peculiar feature is that, as we move from the left to the right side of the equivalence sign, 'we carve up the content in a way different from the original one and this yields us a new concept' (*ibid.*, p. 75).

The *Foundations* does not explain this puzzling remark; we must wait till the essay 'Function and Concept' to gain further illumination. Frege does not pursue the issue in his book because he has other grounds for rejecting (4) and (5) as possible definitions of direction and number respectively. Such definitions, Frege argues, do not specify the truth-conditions of identities such as

(6) $D(a) = q$

and

(7) $N_xFx = q$

where q is neither a direction nor a number. Under this condition the two propositions should be false, but that conclusion cannot be justified on the basis of the proposed definitions (4) and (5). In order to show the falsity of (6) and (7) we require the general notion of direction and that of number respectively. If we had them, we could determine that, if q is not a direction or is not a number, then (6) and (7) are false. But definitions of the form (4) and (5) do not provide us with those two general notions. Frege concludes reluctantly that his originally proposed definitions must be abandoned in favor of the explicit definitions

(8) $D(a) =$ the extension of $(x//a)$

and

(9) $N_xFx =$ the extension of (equinumerous with Fx)

Unfortunately, definition (9) does not show why the notion of number is a logical notion. It is expressed in terms of the extension of a concept, but there is no argument in the *Foundations* to show that extensions of concepts are logical objects. In 'Function and Concept' this lacuna is supposed to be filled by showing that extensions are value-ranges and value-ranges are characterized by something like proposition (1). That basic proposition characterizes value-ranges both contextually and by laying down identity conditions for them. It is structurally similar to the equivalences (4) and (5), but there are two reasons why Frege thinks it escapes the objections raised against the equivalences as definitions. First, (1) is taken as an axiom rather than as a definition, and, second, Frege hopes to show in the *Grundgesetze* that (1) uniquely determines value-ranges within the context of his formal system (cf. chapter VI).

There still remains the question why proposition (1) should be considered a truth of logic. At this point Frege has to turn back to the idea in the *Foundations* that the constituent propositions of (4) and (5) carve up the same content in different ways and thus yield new concepts. Presumably the same thing is to hold of the constituent propositions of (1).

It is clearly insufficient to assume that proposition (1) is logically true because the sentences on both sides of the equivalence sign have the same content in the *Begriffsschrift* sense. That would only mean that they designate the same thing or, to use the later terminology, that they have the same reference.[16] It is also not possible to say that the two sentences have the same mode of designation or form. Clearly they do not; the two sentences are said to carve up the content in different ways and to introduce new concepts. In order to explain what makes proposition (1) logically true, we want to be able to say that the two constituent propositions have the same meaning, where that means neither the same form nor the same content. A third intermediate level must be postulated and that is what Frege calls the level of sense. What is the same in the propositions is not their linguistic form, but their cognitive content. The sentence on the left side of the equivalence sign 'expresses the same sense . . . but in a different way. It represents the sense as the generalization of an equation, whereas the newly introduced expression is simply an equation whose left side as well as its right side has a complete meaning' (KS, p. 130).

This is clearly not intended as a proof that proposition (1) is a truth of logic. Frege after all calls it a basic law of logic. It merely shows why Frege considered it logically true: he regards the two sentences as having the same sense and their equivalence can therefore be known *a priori*. The proposition is analytic in precisely the Kantian sense. By analysing what is on one side of the equivalence sign we obtain what is on the other side without recourse to an external object. It is Kantian in the sense that this analysis is not carried out at the linguistic level, but at the cognitive level.

When we think the thought expressed by the sentence on the left, then we are thinking the thought expressed by the sentence on the right 'though not so clearly and with the same consciousness.'

The conclusion is surprising in that it might seem that the two different sentences connected in (1) are really about different subject matter. One is concerned with functions, the other with value-ranges. We now discover that this is a misleading way to put it. For not only have the two sentences the same content or reference, they even have the same sense. It is only our subjective perception and our manner of speaking that distinguish the statement about functions from that about value-ranges. This conclusion in no way subverts the strict separation of functions and objects. It is not that functions are the same as value-ranges; it is just that a thought concerning a function is the same as one concerning a value-range.

8 Truth as the reference of sentences

In recent years the Fregean doctrine of sense and reference has been the predominant focus of attention in the philosophical study of his work. The essay 'On Sense and Reference' is Frege's most anthologized, most translated, and most widely read piece of work. From the point of view of contemporary analytic philosophy the doctrine may look like his most distinctive and most interesting idea.

There are a number of reasons for this way of reading Frege. His logical achievements have been largely absorbed into contemporary logic. His philosophy of mathematics was blown apart by the discovery of Russell's antinomy. What remains distinctive in his work is therefore thought to be his contribution to philosophical logic, at the center of which is the theory of sense and reference. For that reason Dummett can call Frege's philosophy of mathematics 'a starting-point only in a historical sense,'[17] while maintaining that his work in philosophical logic 'is a true foundation' that must serve 'as the starting-point for anyone working in this area today' and that provides the terms 'in which the basic problems can still most fruitfully be posed' (*ibid.*, p. xiv).

While it is true that the doctrine of sense and reference and the issues it raises have been a major concern of recent analytic philosophy of language, it is important to keep in mind that Frege's interest in that doctrine differs from the current one in several respects. He introduces the theory initially to explain why the axiom of value-ranges should be considered a logical truth. The requirements of his philosophy of mathematics are what motivate the introduction of epistemological considerations into his theory of meaning. Once the value-range axiom was abandoned as a result of Russell's antinomy the tie he saw between the philosophy of mathematics and the philosophy of language was severed.

Frege's other interest in the distinction of sense and reference is also different from that of contemporary philosophy of language. In the *Grundgesetze* he singles out the introduction of truth-values as the most significant achievement of the theory of reference. This assessment is reflected in the essay 'On Sense and Reference,' where two-thirds of the text is concerned with explaining how the distinction applies to sentences. After it has been argued that the reference of a sentence is a truth-value, the remainder of the essay tries to show how apparent counter-examples to the claim can be accounted for.

There are good reasons why the question of the reference of sentences should have been such a major issue to Frege. In the *Begriffsschrift* he had constructed a truth-functional logic; later on he had argued that logic is concerned with the laws of truth; and he had tried to defend an extensional view of concepts. In order to explain these features of his logic it was not sufficient to distinguish between a thought and its truth and falsity. That distinction had been made long before the discovery of sense and reference (cf. NS, p. 8). The progress represented by the doctrine of sense and reference is that it assigns thoughts to a different semantic level from that of truth-values. With that distinction Frege can then assume that the truth-functional logic of the *Begriffsschrift* does not merely describe an arbitrary fragment of the logical relations between judgments, but is complete and self-contained. The sense in which logic can be said to deal with the laws of truth becomes more precise and the extensional view of concepts is philosophically supported.

Once it is assumed that Frege's doctrine is meant primarily as a theory about referring expressions, that erroneous interpretation draws some spurious confirmation from the now-entrenched translation of Frege's term '*Bedeutung*' as 'reference.' Ernst Tugendhat[18] was the first to notice this fact:

> The rendering in English of Frege's term '*Bedeutung*' as 'reference' . . . is quite as misleading as the earlier renderings 'denotation' and 'nominatum.' . . . The translators have preferred to withhold from English readers the puzzlement which every German reader experiences with this word on first reading Frege's essay '*Über Sinn und Bedeutung.*' They chose to anticipate an answer, and to have done this is perhaps worse than that it happens to be the wrong one.

To avoid any misleading connotations Tugendhat suggests the term 'significance' as an alternative translation of '*Bedeutung.*'[19] Instead of adopting that suggestion, it seems wiser to retain the established translation of the term since a shift in terminology at this point would only engender new confusions. Nevertheless, Tugendhat is surely right in the substantial point he is making. It is that the semantics of sense and reference is primarily a semantics of whole sentences and not of sentence parts. In support of his thesis he quotes a revealing passage from the essay 'On Sense and Reference' where Frege writes:[20]

> That we are concerned at all with the reference of a sentence part is an indication that, in general, we recognize and demand a reference for the sentence itself. . . . Why do we require that every proper name should have a reference and not merely a sense? Why are we not satisfied with the thought? Because, and in so far as, we are concerned with its truth-value. . . . The striving for truth is therefore what drives us everywhere from a sense to a reference (KS, p. 149).

The claim that after 1891 the name/bearer relationship is the paradigm of Frege's semantics and that his theory of sense and reference is primarily meant as a theory of referring expressions has the effect of assigning a basic role to empirical objects. But it seems doubtful that such objects could ever have played an important role in Frege's thought. He does not regard empirical objects as items of acquaintance that can be simply named or described. 'Observation itself already includes a logical activity' (F, p. 99). It had been one of Frege's assumptions against physiologically oriented psychologism that sensation never presents us with material objects. In the late essay 'The Thought' he argues that sensory impressions are necessary but not sufficient for seeing things. Something non-sensory must be added to the impressions and only with that addition do we gain access to the empirical world with its empirical objects (KS, p. 360).

Frege's mode of exposition in the essay 'On Sense and Reference' may seem at first sight to lend plausibility to the idea that the doctrine is intended primarily as a theory of names and definite descriptions. The paradigm of reference seems to be the relation between a name and what it names. In the first paragraphs of the paper Frege seems to use the distinction between informative and non-informative identity statements to draw attention to the need for distinguishing sense and reference of names. 'It is natural, now, to think of there being connected with a sign (name, combination of words, letter), not only that to which the sign refers . . . but also what I should like to call the *sense* of the sign' (KS, p. 144). After that Frege seems entirely occupied with a discussion of proper names and definite descriptions until the second part of the essay, where he suddenly raises new and seemingly unrelated questions about 'the sense and reference for an entire declarative sentence' (*ibid.*, p. 148). The rest of the essay is taken up with the examination of the distinction of sense and reference for sentences. It appears from this that Frege began with names and then moved on to sentences, thus assimilating sentences to names.

However, closer examination of the text reveals that this interpretation is unsatisfactory. In the introductory paragraphs Frege raises a problem about identity statements. He asks how it is possible that two sentences can deal with the same subject matter and have the same truth-value while none the less differing in epistemological value. 'The evening star is the evening star' and 'The evening star is the morning star' both deal with the planet Venus and both are true, but one of the sentences is trivial and the other is not. The problem is resolved by distinguishing between the sense and the reference of an expression. That presumably means that we must distinguish between the sense and the reference of the expressions 'the evening star' and 'the morning star.' But implicit in the problem is the assumption that on the basis of this difference we must also distinguish between the sense and the reference of the sentences in which they occur.

When Frege says that the distinction is initially to be applied to 'names' it is necessary to recall that he uses the term 'name' in a peculiar sense. In the terminology previously adopted in 'Function and Concept' a 'name' is not merely a referring expression, but any kind of expression that is not incomplete, i.e., any kind of expression that is not a functional one. An object, in his equally wide use of that term, is anything that is not a function (KS, p. 134). And a (proper) name is anything that is a sign for an object, according to the essay 'On Concept and Object' (*ibid.*, p. 171). Since truth-values are not functions, and sentences are said to refer to truth-values, sentences are therefore names and the sense–reference distinction must apply to sentences. When Frege says in 'On Sense and Reference' that he wishes to draw the distinction of sense and reference initially only for names, he is not thereby excluding sentences from the discussion, but only functional expressions. That is shown in the comment:

> It is clear from the context that by 'sign' and 'name' I have here understood any designation representing a proper name, which thus has as its reference a definite object (this word taken in its widest sense), but not a concept or a relation, which shall be discussed further in another article (KS, p. 144).[21]

When Frege moves in the essay from the exemplification of the distinction between sense and reference for referring expressions to the sense and reference of sentences he tries to make that move as plausible as he can, but he does not give a compelling reason for saying that thoughts are the senses of sentences and truth-values are their references. By arguing that the sense of a sentence must be determined by the senses of its constituents and that the reference of the sentence must likewise be determined by the references of the constituents, he merely succeeds in showing that the thought expressed by a sentence cannot be its reference but might be its sense, and that the truth-value might be the reference of the sentence but cannot be its sense. In other words, from the assumptions concerning the sense and reference of referring expressions he cannot show that, for a sentence, the thought must be the sense and the truth-value must be the reference.

The reason for the inconclusiveness of the argument is not difficult to find. The sense and the reference of a sentence are determined by the sense and the reference of all the constituents of a sentence. But so far we have only been given some indication of the sense and reference of one type of sentence constituent. In order to complete the argument we would have to know what the sense and reference of concept expressions are. But to know that we would have to know what values a concept correlates with given arguments and that in turn would require that we know the reference of a sentence in which the concept expression occurs. Therefore, given Frege's assumptions, any argument designed to prove that sentences refer to truth-values would presuppose that very conclusion.

There is another reason for regarding the exposition in the essay 'On Sense and Reference' as didactic rather than systematic. Not everything that looks like a word is for Frege a meaningful constituent of a sentence. The sense and reference of a sentence are determined by the sense and reference of the constituents of the sentence. But only what contributes to the meaning — the sense and the reference of the whole sentence — counts as a constituent of it.

The distinctive feature of Frege's theory of sense and reference is that it introduces epistemological issues into strictly semantic considerations. In his book *Frege. The Philosophy of Language* Dummett has pursued this bonding of semantics and epistemology further than anyone else. As a result he has tried to recast a number of characteristically epistemological problems and concepts as problems or concepts in the theory of meaning. Notions such as understanding, knowledge, verification, and falsification become for him integral notions of semantics. The dispute between realism and anti-realism is reconstructed as a dispute about the proper theory of meaning. But even if Dummett's arguments show how Frege's idea of bonding semantic and epistemological issues can be developed further, they do not necessarily represent Frege's own thoughts.

His considerations do possibly serve another important function, however. For the complexity and opacity of Dummett's theory might lead one to question the introduction of epistemological considerations into semantics at all. Instead one might prefer a sharp separation of the two sorts of issue. The outlines of such a de-epistemologized semantics can be found in Saul Kripke's account of proper names. By separating the question of the meaning of a referring expression from the way in which the reference of the expression has come to be fixed, Kripke in effect separates the notion of sense from the theory of meaning and speaks of meaning in the way in which the *Begriffsschrift* speaks of the mode of designation.[22]

VI

The End of Logicism

1 Russell's contradiction

The years between 1879 and 1902 were the most creative in Frege's life, but they did not bring him the recognition he longed for. As both philosophical and mathematical journals rejected his essays, he began to suspect that they were too philosophical for most mathematicians and too mathematical for most philosophers (cf. WB, pp. 1, 134–5, 254, 258). When his writings appeared in print it was often only because he paid for the cost of publication out of his own pocket (cf. WB, pp. 138–9). Even then the response to his ideas was limited. Lacking professional recognition, he took seventeen years to be promoted to *ausserordentlicher Professor*.

When the publication of the first volume of the *Grundgesetze* again met with indifference he began to feel embittered and voiced his bitterness in sharp attacks on those whose views he considered inferior to his own. In 1894 he tore apart Husserl's *Philosophie der Arithmetik*. The following year he turned to the critique of Schröder's main work. In 1896 he unfavorably compared Peano's logic to his own. And finally in 1899 he published a long satirical piece on the first installment of the new *Encyclopedia of the Mathematical Sciences*, in which a schoolmaster by the name of Schubert had examined the foundations of arithmetic. Sarcastically Frege congratulated Schubert for having discovered the importance of non-thinking in science.

In the second volume of the *Grundgesetze* the attack was continued with sharp criticisms of the views of Cantor, Dedekind, Weyerstrass, Heine, and Thomae. Frege's harsh and unsympathetic comments were bound to bring him into conflict with his mathematical colleagues. Of Helmholtz's famous essay 'Zählen und Messen erkenntnistheoretisch betrachtet' he wrote that it attributed magical power to signs and that its reliance on psychology and empiricism helped to make it completely unclear. 'Hardly have I seen anything less philosophical than this philosophical essay and hardly ever has the meaning of the epistemological question been more misunderstood' (GG, vol. 2, p. 140).

THE END OF LOGICISM

On the eve of the publication of the second volume of the *Grundgesetze* in 1902 Frege received a letter from Bertrand Russell that, ironically, signalled both the beginning of Frege's recognition and the end of his creative work. Russell wrote in his letter: 'In many individual questions I find in your writings discussions, distinctions, and definitions for which one looks in vain in the work of other logicians. . . . I am about to complete a book on the principles of mathematics and I would like to give a detailed account of your work in it' (WB, p. 211). Russell fulfilled this promise in the first appendix to his *Principles of Mathematics* in which he gave a lengthy, if idiosyncratic, description of Frege's logical and arithmetical doctrines. When Wittgenstein later read Russell's book the appendix drew his attention to Frege's views and thus initiated the philosophical connection between Frege and Wittgenstein.

In his letter of June 16, 1902, Russell added to his positive assessment of Frege's doctrines the remark:

> Only in one point have I encountered a difficulty . . . because of the following contradiction: let w be the predicate to be a predicate that cannot be predicated of itself. Can w be predicated of itself? Every answer implies the opposite. Therefore one must conclude that w is not a predicate. For the same reason there is no class (as a whole) of those classes as wholes that do not contain themselves. From this I conclude that under certain circumstances a definable set does not form a whole (*ibid.*).

Russell recognized that his discovery demanded major revisions in his own account of predicates and classes. Much of the *Principles of Mathematics* had been written before he made his discovery. The contradiction forced him to revise his book substantially and even so he had to conclude: 'What the complete solution of the difficulty may be, I have not succeeded in discovering; but as it affects the very foundations of reasoning, I earnestly commend the study of it to the attention of all students of logic.'[1]

It took Russell little effort to draw Frege's attention to the significance of his discovery. At the time Russell may not yet have fully understood how seriously his discovery threatened Frege's logicist program. But Frege himself was in no doubt about the potentially deadly force of the contradiction and his response to Russell's letter was forceful and immediate:

> Your discovery of the contradiction has caused me the greatest surprise and, I would almost say, consternation, since it has shaken the basis on which I intended to build arithmetic. . . . The second volume of my *Grundgesetze* is to appear shortly. I shall no doubt have to add an appendix in which your discovery is taken into account. If only I had the right point of view for that (*ibid.*, p. 213).

A modification of the value-range axiom was called for that would preserve the essentials of the system. But how was that modification to be made? The only consolation was that the difficulty also affected other mathematical and logical theories. In September of 1902 Frege wrote to Jourdain:

> Through a letter from Mr. Bertrand Russell my attention has been drawn to the fact that my basic law V requires a restriction. But our correspondence about this point has not yet led to a satisfactory conclusion. By the way, the difficulty is not peculiar to my logic, but similarly occurs in Peano's (*ibid.*, p. 111).

Russell thought that the contradiction could be constructed in more than one form. In his letter to Frege he had first expressed it in terms of predicates and then in terms of classes. In a subsequent letter he added a third version, one involving the notion of proposition (cf. WB, p. 230; also *The Principles of Mathematics*, p. 527). Given the assumptions of his logic, the contradiction appeared serious in each of those forms. For Frege the significance of Russell's discovery was quite different. The problem seemed to him to be due to the way in which he had introduced logical objects.

In reply to Russell's first letter Frege pointed out that the predicative version of the contradiction could not be reconstructed in his own logic. Functions could never be meaningfully arguments of themselves; predicates could never be meaningfully predicated of themselves. The distinction of levels of functions prevented that kind of construction. But the situation was different for the class antinomy. Classes were value-ranges for Frege; value-ranges were objects; and objects are not stratified into different levels. Hence, the question whether a class is or is not a member of itself was a meaningful question. The problem, as he had always feared, was with the value-ranges. He wrote to Russell:

> I myself fought against the recognition of value-ranges and thereby of classes for a long time; but I did not see any other possibility for giving logical foundations to arithmetic. The question is: how do we grasp logical objects? And I have not found any answer but the following: we grasp them as extensions of concepts or, more generally, as value-ranges of functions. I have never been unaware that this leads to difficulties and these have been multiplied by your discovery of the contradiction. But what other way is there? (WB, p. 223).

In the *Grundgesetze* his attitude had been quite different. In the first volume he had written: 'The introduction of the symbolism for value-ranges seems to me one of the most significant additions to my logic' (GG, vol. 1, pp. 15–16). It is true that he had allowed that disagreements might erupt over his value-range axiom (*ibid.*, p. vii), but he had concluded the preface with the confident words: 'I would consider it a refutation only if somebody showed in practice that a better and more solid structure could be erected on different foundations or if somebody proved that my axioms lead to obviously false conclusions. But nobody will be able to do that' (*ibid.*, p. xxvi).

Nevertheless it is not altogether untrue to say that Frege had fought against the recognition of value-ranges and thereby of classes. They had been a later addition to his logic and even in the *Foundations of Arithmetic* he had toyed with the idea that numbers might be definable without making use of the notion of the extension of a concept. Having introduced numbers as extensions, he says in the *Foundations* that the term 'the extension of the concept' might be replaceable by the term 'the concept.' And he adds:

> But this would be open to the two objections:
> 1. that this contradicts my earlier statement that the individual numbers are objects, . . .
> 2. that concepts can have identical extensions without themselves coinciding. As it happens, I am convinced that both these objections can be resolved; but to do so would lead us too far afield here (F, p. 80).

In his later writings Frege never explicitly came back to this passage. When Benno Kerry asked him about its meaning in 1887 he replied somewhat brusquely that he had built nothing on it (KS, p. 172).[2]

One year after the publication of the *Foundations* Frege wrote of his logical analysis of the numbers: 'I have replaced the expression "set," which is frequently used by mathematicians, by "concept," which is usual in logic' (KS, p. 105). It is not altogether clear that this remark implies the possibility of a definition of numbers which does not make use of the notion of the extension of a concept, but in any case Heinrich Scholz reports that in a manuscript written after 1884 Frege 'attempts in particular to define "the number which belongs to a concept" without using extensions of concepts.'[3]

Frege was therefore not misrepresenting the facts when he wrote to Russell in 1902 that he had always felt uneasy about value-ranges. Nevertheless, he had eventually concluded that there was no other way to get to logical objects. And thus he had felt justified in introducing them into his logical system.

2 *Frege's way out*

In the system of the *Grundgesetze* value-ranges are not definable. They are introduced and characterized by axiom V of the system, which can be written as:

$$(1) \quad \forall x\, (fx = gx) \leftrightarrow (\acute{x} fx = \acute{x} gx).$$

THE END OF LOGICISM

There is some question whether this axiom in fact gives a unique characterization of value-ranges. It seems that it does not. For given a function which fulfills the following two conditions:

$$(2) \quad x = y \leftrightarrow h(x) = h(y)$$

$$(3) \quad x \neq h(x)$$

it follows clearly from (1) that

$$(4) \quad \forall x (fx = gx) \leftrightarrow h(\acute{x}fx) = h(\acute{x}gx).$$

In other words,

$$(5) \quad \forall x (fx = gx)$$

is a necessary and sufficient condition for

$$(6) \quad \acute{x}fx = \acute{x}gx$$

and for

$$(7) \quad h(\acute{x}fx) = h(\acute{x}gx).$$

On the other hand, the value-range of a function is never identical with the value of h for that value-range:

$$(8) \quad \acute{x}fx \neq h(\acute{x}fx).$$

Frege argues that this indeterminateness is overcome when every function is introduced in such a way that its value for any possible argument, including any possible value-range, is determined (GG, vol. 1, p. 16). According to the principle of the identity of indiscernibles two objects are identical when every predicate that applies to the one also applies to the other and vice versa. In order to uniquely determine the identity of value-ranges, Frege considers it sufficient to determine for every function what the value of that function is for a given value-range.

That means also that we must determine the truth-value of the identity

$$(9) \quad \acute{x}fx = q$$

where 'q' stands for something which is not a value-range. In the system of the *Grundgesetze* this problem can be resolved by noticing that apart from value-ranges themselves the only objects that occur are truth-values. Frege shows that in his system truth-values can be identified with their own unit sets and that therefore the question of the truth conditions for (9) can be reduced to the question of the truth conditions for an identity of the form of (6) and that truth condition is determined by (1). He concludes that 'we have thus determined the value-ranges in so far as that is possible here' (*ibid.*, p. 18). If new functions are to be introduced later we must determine what their values are for value-ranges as arguments 'and this may be considered as a determination of the value-ranges just as much as it may be considered a determination of those functions' (*ibid.*).

Given the way in which he has introduced his elementary symbols, Frege thinks he can prove that every such symbol has exactly one reference (*ibid.*, pp. 45ff). In particular he thinks he has ensured that well-formed expressions of the form '$\acute{x}fx$' have a unique reference. Thereby he hopes to have established that 'the same also holds of any properly formed complex name' (*ibid.*, p. 50). Since he treats sentences as names of truth-values, the proof that every name has exactly one reference amounts to a proof that no sentence refers to both the True and the False. If Frege's argument were successful it would in effect give us a consistency proof for the system of the *Grundgesetze*.[4]

Frege's attempt to show that his basic law V uniquely determines value-ranges in the system of the *Grundgesetze* depends on the assumption that every function should have a determinate value for every possible argument and that in particular every first-level function should have such a value for any possible object as argument. Frege considers this requirement both integral to the formal system and compelling on philosophical grounds. He writes:

> One can express it metaphorically thus: the concept must have a sharp boundary. When one compares concepts with respect to their extension with areas of a plane, we have a metaphor that one can use only with caution, but that can be of good service here. To a concept without a sharp boundary there would correspond an area whose borderline is not sharp everywhere and which in places blends imperceptibly into its surroundings. Strictly speaking, that would not be an area and similarly it would be a mistake to consider a concept without sharp boundaries a concept. Logic cannot recognize such concept-like formations as concepts; it is impossible to form precise laws for them. The law of excluded middle is really nothing but another form of the demand that the concept should have a sharp boundary (GG, vol. 2, p. 69).

These considerations may seem completely plausible, but in Frege's system they lead inevitably to Russell's contradiction. Numbers are, for Frege, objects. These objects can be found located in the value-ranges. Value-ranges as logical objects must be introduced through an axiom that lays down identity conditions for them in such a way that reference is made only to universal characteristics of functions. In order for axiom V to fulfill that condition we must assume that every function of the system of the *Grundgesetze* is completely defined.

If c is a class, the function x *is not an element of* c must be defined for every object and that means for c itself. In other words, 'c is not an element of c' must be either true or false. It follows that the concept x *is not an element of* x must be well defined. Axiom V assures us that for every concept there exists exactly one corresponding class. Let S be the class corresponding to the concept x *is not an element of* x. The concept x *is an element of* S must be defined for every object including the class S. If we try to determine whether 'S is an element of S' is true or false, we discover that S is an element of S if and only if S is not an element of itself.

The contradiction seems to make it necessary to abandon either the assumption that value-ranges are objects or the demand that every first-level function should be defined for any possible object. To abandon either would be to abandon an integral element of Frege's account of value-ranges as logical objects and would thus seem to undermine the very possibility of a logical analysis of arithmetic. For this reason Frege could write that with Russell's discovery 'not only the foundations of my arithmetic, but also the sole possible foundations of arithmetic seem to vanish' (WB, p. 213).

In spite of the setback Frege did not abandon his project in despair. With what Russell has called an almost superhuman effort,[5] Frege set out to find a positive aspect in Russell's discovery. In his letter to Russell he wrote:

> Yet, I should think that it must be possible to set up conditions for the transformation of a generalized equality into an equality of value-ranges so that the essentials of my proof remain intact. In any case your discovery is very remarkable and will perhaps result in a great advance in logic, unwelcome as it may seem at first glance (WB, p. 213).

In the appendix to the second volume of the *Grundgesetze* Frege tries to show how those essentials could be salvaged. The appendix explains how the contradiction arises from the original value-range axiom, reviews the various possibilities for avoiding the contradiction which he had discussed with Russell in the few months since the letter of June 16th, and finally suggests his own amendment to the axiom. The original axiom is equivalent to

$$(10) \quad \forall y \, (y \in \acute{x} Fx \leftrightarrow Fy)$$

If we let Fy be $-(y \epsilon y)$, we obtain the contradiction by universal instantiation:

$$(11) \quad (\acute{y} - (y \epsilon y) \in \acute{y} - (y \epsilon y)) \leftrightarrow -(\acute{y} - (y \epsilon y) \in \acute{y} - (y \epsilon y))$$

This inference is blocked if we replace (10) by

$$(12) \quad \forall y((y \in \acute{x} Fx) \leftrightarrow (Fy \,\&\, y \neq \acute{x} Fx))$$

The value-range axiom can then be modified appropriately to become logically equivalent to (12).

The fundamental problem is how to grasp numbers as logical objects. The appendix and the whole book conclude with the hopeful words: 'Even if this problem has not been resolved to the extent to which I considered it resolved at the time I was writing this volume, I do not doubt that the road to the solution has been found' (GG, vol. 2, p. 265).

As the years went by Frege's confidence in the proposed solution seems to have weakened. He never explicitly repudiated it, but he also never pursued it any further. Eventually, his doubts began to extend to the notion of set or value-range itself. In 1906 he set out to write on the logical paradoxes of set theory. The outline of the uncompleted text shows that he had planned to discuss the solution to the problem which he had suggested in the appendix of the *Grundgesetze*. But this solution was not to be the final word of the essay. The outline ends with the words: 'set theory shaken. My logic on the whole independent of this (in contrast to other similar projects)' (NS, p. 191).

Frege was right in feeling dissatisfied with the solution to the contradiction that he had proposed in the appendix to the *Grundgesetze*. The amended axiom blocks the derivation of Russell's antinomy, but it is neither formally nor philosophically satisfactory. It is formally unsatisfactory because new contradictions can be derived from it, together with the assumption that there are at least two objects.[6] That was surely an assumption to which Frege was committed, since he assumed truth-values to be objects and there were two truth-values. Quite apart from that difficulty, the new axiom would not allow for the logical derivation of arithmetic. There is no evidence that Frege ever came to see that the amended axiom was insufficient for deriving the principles of arithmetic, but (as Dummett has suggested) it seems plausible to assume that he checked the proofs in the *Grundgesetze* and discovered that the substitution for the original value-range axiom could not perform the tasks of the old one. In any case, the projected continuation of the *Grundgesetze* never materialized and the solution of the appendix to the second volume was never systematically pursued.

Besides the technical difficulties there may have been philosophical grounds for Frege's dissatisfaction. The original value-range axiom had been chosen carefully on the basis of reflections about the nature of numbers, logic, and objects. Whatever intuitive justification the original axiom might have had was certainly missing from the modification. That might have been acceptable if Frege's aim had been merely to formulate a consistent theory of classes from which the normal arithmetical principles could be derived, but his purpose was deeper and more philosophical. He had hoped to show that arithmetical laws are analytic and therefore *a priori* truths and that empiricism is a deficient philosophy. How could the derivation of arithmetical principles from purely pragmatically adopted axioms support that point?

It took him time to adjust to the situation. By 1906 he was beginning to think that the theory of sets was undermined by the contradiction. He concluded that there was no use for sets or classes anymore. Much later he wrote:

> There is a tendency in language, disastrous for the reliability of thinking, to create proper names to which no objects correspond.... A particularly noteworthy example is the formation of a proper name on the pattern of 'the extension of the concept a' as, for instance, 'the extension of the concept fixed star.' This expression seems to refer to an object because of its definite article, but there is no object that could be referred to in this way in language. From this the antinomies of set theory have arisen. I myself was fooled by this deceptive appearance when I attempted to give logical foundations to the numbers by trying to conceive them as sets (NS, pp. 288–9).

By 1919 he was wondering whether numbers were objects at all. As he had previously done in the *Foundations of Arithmetic* he argued that one could quite consistently define second-level concepts and relations which somehow reflected the laws of arithmetic.

> But in them we do not have the numbers of arithmetic; we do not have objects, but only concepts. How is it possible to proceed in an unobjectionable manner from those concepts to the numbers of arithmetic? Or are there no such things as the numbers of arithmetic? Are the number terms dependent parts of the signs for those second-level concepts? (NS, p. 277).

These questions conclude the summary of his scientific and philosophical achievements for Darmstaedter in 1919. What is most striking is that the summary makes no mention at all of value-ranges or classes. By 1919 Frege had given up all hope of grounding arithmetic in that way. 'When one considers these questions at length,' he noted in his diary in 1924, 'one begins to suspect that linguistic usage is misleading and that number terms are not proper names of objects at all' (NS, p. 282).

3 The search for new foundations

When Rudolf Carnap[7] attended Frege's lectures at Jena in 1910 he found that

> Frege looked old beyond his years. He was of small stature, rather shy, extremely introverted. He seldom looked at the audience. Ordinarily we saw only his back, while he drew the strange diagrams of his symbolism on the blackboard and explained them. Never did a student ask a question or make a remark whether during the lecture or afterwards. The possibility of discussion seemed out of the question.

There is no doubt that Frege had always been an introverted personality, but the experience of the collapse of his long-term project must have contributed to his withdrawal from the surrounding world. In the lectures Carnap attended Frege gave no indication of how to deal with the problems created by Russell's discovery. In 1911 he suggested that he still believed the new logic could serve for the construction of the whole of mathematics, but neither in the lectures of 1911 nor in those of 1913 did he show how such a construction could be brought about. And in neither lecture course did he refer to the notion of the extension of a concept.[8]

Nevertheless, Frege had not abandoned his project in despair. As he became less certain of the attempt to reduce arithmetic to logic, his attention was naturally drawn back to geometry. Kant had argued that there must be *a priori* knowledge because both arithmetical and geometrical truths are synthetic *a priori*. In contrast Frege had subscribed to the Lotzean doctrine that arithmetical truths are analytic and had thought that this fact might provide a stronger argument for the existence of *a priori* knowledge than the claim that geometry is synthetic and *a priori*. None the less he had agreed with the Kantian assessment of the status of geometry. As the belief in the analyticity of arithmetic became more doubtful, the possibility of relying on the synthetic *a priori* character of geometry to show that human knowledge has *a priori* foundations became correspondingly more attractive.

Frege's long-standing interest in geometry had been revived shortly before he learnt of Russell's contradiction by the publication of the first version of Hilbert's *Grundlagen der Geometrie* in 1899. Frege then wrote to tell Hilbert that he had taken a special interest in the work 'in particular since I myself have earlier been concerned with the foundations of geometry though I have not published anything on the subject' (WB, p. 60). In the ensuing correspondence between them Frege criticizes Hilbert's characterization of his axioms as explanations or definitions of the geometrical terms contained in them. He also rejects Hilbert's assumption that we can regard the axioms as true when they are proved consistent. In Frege's eyes both claims are incompatible with the assumptions that truth is basic to any theory, including geometry. An axiom is the expression of a recognized truth for him and hence must be distinguished sharply from a definition which is neither true nor false but only practical or impractical. The truth of the axioms cannot be derived from their consistency—only the reverse is possible.

These criticisms are repeated in two series of essays on the foundations of geometry which Frege published in 1903 and 1906 respectively (cf. KS, pp. 262–72 and 281–323). They brought little response from Hilbert. With respect to Frege's second objection Hilbert replied:

You write: 'I call axioms propositions. . . . From the truth of the axioms it follows that they cannot contradict each other.' I found it interesting to read just this sentence in your letter, since I have always maintained the very opposite as long as I have thought, written, and lectured on the matter. When arbitrarily adopted axioms do not contradict each other in any of their consequences then they are true and the things defined by the axioms exist. That is for me the criterion of truth and existence (WB, p. 66).

As far as Frege's first objection was concerned Hilbert had to admit that the axioms do not uniquely specify the concepts that are said to be defined by them. Instead he argued that they specify only a structural network or schema of concepts and the theory can be applied to 'infinitely many systems of basic elements' (*ibid.*, p. 67). This is in fact to grant Frege's point that axioms considered as definitions do not define geometrical concepts themselves but only second-level properties which we assume our geometrical concepts possess. In the later editions of his work Hilbert tried to accommodate this Fregean point.

The exchange between Frege and Hilbert and Frege's essays on the foundations of geometry are disappointing because the interests and views of the two opponents remain too far apart. What separates Frege from Hilbert is not just a disagreement about the role of axioms in geometry, but an altogether different conception of the nature of geometry itself. Hilbert's work is the culmination of a process that began with Gauss's suggestion of the possibility of a non-Euclidean geometry. If we take that suggestion seriously we begin to drain geometry of its intuitive content and turn it into a formal theory. In Hilbert's work the separation of the formal theory from its intuitive filling is complete. Frege, on the other hand, believes with Kant that geometry is a theory of intuitive contents. Just as he rejects the formalist interpretation of arithmetic he also rejects the formalist interpretation of geometry expressed in Hilbert's work.

Unfortunately Frege has almost nothing to say in defence of the claim that geometry requires a synthetic *a priori* basis in intuition—at least not in the writings that have survived. That makes his claims about geometry a good deal less interesting than the proposition that arithmetic is analytic. The latter gave rise to his new logic, to reflections about the nature of logic, logical objects, and logical objectivity, and to an elaborate theory of meaning. Nothing comparable came from his view of geometry. In order to make that view interesting he would have had to develop an account of intuition, to produce arguments for the possibility of pure intuitions and considerations in support of the thesis that space and time are objects of pure intuition. None of those details are present in what he says about geometry.

One thing that is clear about Frege's late concern with geometry is that it comes from the same motivations that originally drove him to defend the thesis that arithmetic consists of analytic truths. In a very late essay on

'The Sources of Knowledge in Mathematics and the Mathematical Sciences' (NS, pp. 286-94), he distinguishes three such sources: sensory experience, logic and geometry (*ibid.*, p. 286). He allows that sensory impressions can cause us to make judgments, but since those judgments may be either true or false, sensory impressions can give rise to error. For that reason 'sensory perception as a source of knowledge should be considered unreliable and of limited value' (*ibid.*, p. 287). The difficult question is how we can distinguish perceptual error from veridical perception. 'If all events were equally lawless or if the laws of physical events were unknowable for us then we would lack any means of recognizing perceptual errors. The natural laws already recognized allow us to avoid sense deceptions' (*ibid.*). Those laws cannot themselves be derived entirely from sense perception. In order to avoid circularity we must assume that natural laws are partly recognized through other sources of knowledge, namely, logic and geometry. 'In this way we can advance step by step: a progress in the recognition of the natural laws saves us from sensory deception and the more and more purified perceptions help us to a better knowledge of the natural laws' (*ibid.*). Perception is therefore necessary for scientific knowledge but not sufficient; other sources of knowledge are also required and only their combination allows us deeper penetration into mathematical physics.

In this essay on the sources of knowledge Frege goes on to say that logic cannot provide us with objects and that the belief that there are extensions of concepts is based on a false reliance on language. Only geometry can provide us with the infinite totality of objects that we require for defining numbers:

> It is obvious that we cannot attain anything infinite through sensory perception. However many stars we may catalogue there will never be infinitely many in our list and it is the same with the grains of sand at the ocean. Wherever we are justified in talking about the infinite we have not reached it through sensory perception. In order to reach it we require a special source of knowledge and that is the geometrical one (*ibid.*, p. 294).

With these words Frege has broken the last tie to the logicist program. The attempt to reduce arithmetic to logic has been rejected as a failure. The foundations of arithmetic are not to be sought in logic at all, but in geometry. Lotze's logicism has therefore been sacrificed so that Kantian apriorism might live. The search for logical objects has ended and Frege is now willing to grant that objects can be given only in intuition, be it empirical or pure.

When Frege had completed the essay on the sources of knowledge he sent it, on Bauch's advice, to the Neo-Kantian philosopher Hönigswald for publication in his series *Wissenschaftliche Grundfragen*. Hönigswald deeply appreciated the Kantian motivations of the essay. 'The spirit that fills it and the methodological results you have achieved correspond completely to our aspirations,' he wrote (WB, p. 84). Frege's reply was to acknowledge 'that we agree very well on the major points.' In the end the essay was not published in the *Wissenschaftliche Grundfragen*, but only because Frege's death prevented its revision and elaboration.

In Frege's earlier work arithmetic and geometry had been strictly separated. At the end of his life he came to consider the separation artificial and misleading. 'The more I have thought about it the more I have become convinced that arithmetic and geometry have grown out of the same root and that this is the geometrical one, so that the whole of mathematics is really geometry' (NS, p. 297).

In the *Foundations of Arithmetic* he had observed that natural numbers are used for counting and had tried to define them as counting numbers. With the attempt to give geometrical foundations to arithmetic that procedure appeared irrelevant. 'The fact that numbers are used for counting has sprung psychologically from the requirements of practical life and has led scholars into error' (*ibid.*). The attempt to define first the natural numbers and then other types of numbers also had to be abandoned. The natural numbers are 'baby numbers' that can deal only with the discrete. 'Thus the usefulness of the baby numbers is strictly limited.' There are other kinds of numbers, 'but there is no bridge that leads from the baby numbers to them. I myself once thought that one could conquer the whole realm of numbers by proceeding in a purely logical way from the baby numbers upwards; but I have come to see my error' (*ibid.*, p. 296).

In a fragment entitled 'Attempt at a New Foundation of Arithmetic' he tries to show how the numbers can be defined geometrically.

> Diverging from the usual procedure I shall not start from the positive integers and then slowly extend the range of what I call number. For it is really a mistake to have no definite meaning for the word 'number' and always to understand something new by it. The fact that this has happened in the course of historical development is really no objection; for we must always try to establish a closed system in mathematics.
> . . . It is for this reason that I proceed immediately to our final goal, the ordinary complex numbers (NS, p. 299).

The complex numbers are to be defined first and the real and natural numbers are to be characterized as subsets of the totality of complex numbers.

The basic notions of the new theory are point, straight line, and the three-place relation of a point being symmetrical to another point relative to a straight line. Given these geometrical concepts we can define the relationship obtaining when two lines A and B have the same ratio as two other lines C and D. Given a plane with a point of origin A and an end point O it can be shown that, to every pair of lines on the plane that have the same ratio, there corresponds exactly one point C on the plane such that the ratio of AC to OC is the same as the ratio of the given pair of lines. In a directed plane every ratio is thus represented by a point C. This point can be regarded as an ordinary complex number. It was Gauss who first discovered that, if real numbers could be correlated with the points on a straight line, complex numbers could be correlated with the points on a plane. In his final thoughts Frege returned to that insight.

There are many unresolved problems in his attempt to provide new foundations for arithmetic. Some of them he might have resolved if death had not intervened, but in retrospect it is not clear that his new line of argument was very promising. If he found himself forced back to the view that mathematics is geometry, it was only because that was as close as he could remain to his original position after Russell's contradiction. With that contradiction logicism in the sense in which Frege understood it was dead; the possibility of logical objects in his sense had been disproved. The only way in which he could salvage the thesis that mathematical truths are *a priori* and that human knowledge therefore has non-empirical foundations was to turn to geometry. Frege was right in thinking that substantial new advances would come out of Russell's discovery of the contradiction, but they were made by Russell in his theory of types, by Gödel in his incompleteness proofs, and by Tarski in his metamathematics—and not by Frege himself.

These formal developments were not the only results of the collapse of Fregean logicism; it also led to new philosophical elements in the emerging analytic tradition. Moore and Russell contributed ontological and epistemological ideas that were in part derived from the British traditions of common-sense philosophy and empiricism. Through Carnap the ideas of Frege and Russell were combined with problems taken from the critical positivism of Mach. Wittgenstein created his own synthesis of Fregean and Russellian notions with doctrines derived from Mauthner and Brouwer. As the analytic tradition grew it became more empiricist, more atomist, and less Kantian. The focus of its interest shifted from the philosophy of mathematics to the methodology of science and from there to the philosophy of natural language and the philosophy of mind. Its conception of the distinctions between *a priori* and empirical knowledge and between analytic and synthetic truth underwent deep changes. Frege would probably have repudiated almost all these later developments. And yet there is an evident continuity that links his work to that of the later analytic tradition. His logic, his analytic methodology, his formalism and apriorism, his rejection of radical empiricism, his anti-psychologism, and his thoroughly unhistorical view of meaning all are reflected in the later stages of the tradition.

4 Russell

It is common to regard Moore and Russell rather than Frege as the beginning of the analytic tradition, but that leaves Frege's relation to the tradition in limbo. It is more natural to consider Frege as one of the origins of the movement and Moore and Russell as another. By acknowledging two different sources we are in a better position to understand the peculiar combination of German and British, rationalist and empiricist, elements that characterize the whole tradition from its beginning to the present day.

Moore and Russell were born a generation after Frege and brought up in a philosphical climate very different from that of Germany in the last quarter of the nineteenth century. They grew up with a prevailing tendency 'in the direction of German idealism, either Kantian or Hegelian.'[9] Of these two influences the Hegelian was dominant. When Moore and Russell shed their youthful idealism their main target was Bradley's version of Hegelianism, though Russell also later characterized his *Principles of Mathematics* of 1903 as 'a parenthesis in the refutation of Kant.'[10]

Their revolt against idealism initially took the form of an extreme realism. 'I began to believe everything the Hegelians disbelieved,' Russell remembered. 'This gave me a very full universe. I imagined all the numbers sitting in a row in a Platonic heaven. . . . I thought that points in space and instants of time were actually existing entities. . . . I believed in a world of universals. . . . As time went on my universe became less luxuriant.'[11]

Analytic philosophers who have considered Frege in relation to Moore and Russell have tended to assume that their philosophical motivations were more or less identical. This tendency is particularly clear in Dummett's assertion that Frege was a realist in revolt against a dominant Hegelianism. But, unlike Russell and Moore, Frege was in no way concerned with bringing down Hegelianism and if he was committed to a form of realism at all it was of a completely different kind from Russell's or Moore's.

Frege's so-called realism was deeply embedded in the presuppositions of Kant's and Lotze's philosophy. And through that link it stayed free of the epistemological and ontological atomism which became an integral part of Moore's and Russell's realism. When they rejected idealism they came to believe that every word in a meaningful sentence stands for a concept or term and that every such concept or term is real. 'A concept is not in any intelligible sense an "adjective," as if there were something more substantive and more ultimate than it,' Moore wrote in his early anti-idealist essay 'The Nature of Judgment.'[12] Russell echoed this sentiment in the *Principles of Mathematics*: 'Every word occurring in a sentence must have some meaning. . . . Whatever may be an object of thought, or may occur in any true or false proposition, or can be counted as one, I call a term. . . . Every term is immutable and indestructible' (pp. 42ff). And every such term has being (*ibid.*, p. 449).

For Russell we have knowledge by immediate acquaintance with simple concepts or terms. Such entities can be perceived (*ibid.*, p. xv). Even non-spatio-temporal entities 'are in some sense perceived . . . their relation also must be in part immediately apprehended' (*ibid.*, p. 129). Concepts or terms are combined in judgments or propositions, where they form a unity. But in every proposition every term must have meaning and being on its own. This doctrine, which is of course not at all empiricist, shares with traditional empiricism an atomistic and compositional view of judgments.

At the same time as Russell was becoming disaffected with idealism he was also discovering an interest in logic. That interest had been aroused by a study of Leibniz on which he had embarked more or less by accident. It was stimulated by the lecture on symbolic logic that Peano delivered at the International Congress of Philosophy in Paris in 1900. With his usual enthusiasm Russell studied the works of Peano, convinced himself that mathematics was only an extension of logic, and within little more than a year was engaged in writing a book on the subject. In 1903 it was ready for publication under the title *The Principles of Mathematics*.

Philosophically the book relied heavily on the ideas that were worked out by Moore in his essay 'The Nature of Judgment.' And it was the reliance on Moore's extreme realism that caused the greatest difficulties of the book. One problem was connected with the analysis of general propositions. In accordance with Moore's assertion that every term in a judgment has being, Russell argued that terms like 'all men,' 'some man,' 'any man,' and 'the man' each represent an independent concept. In order to construct a satisfactory account of the logical behaviour of general propositions Russell found it necessary to invent an elaborate theory of denoting complexes, to which he devoted a long and difficult chapter of his book.[13]

The other major problem of the book was raised by the contradictions Russell had communicated to Frege. They were at least as serious for him as they were for Frege, since Russell had argued that the doctrine that every term in a judgment has being meant that every such term could be the subject of any possible predicate. If a concept had real being it would have to be predicable of itself and if a class was really one thing it would have to be able to be a member of itself. But the contradiction showed this to be impossible.

Russell communicated his strange discovery to Peano, who seems not to have taken it seriously. In any case he did not respond to Russell's letter (WB, p. 212). From Peano's writings Russell had learnt that Frege was also interested in the construction of a new logic and in the logical derivation of arithmetic from logic.[14] When he received no reply from Peano he naturally turned to Frege. Frege's response was immediate; the precise construction of his logic allowed him to see instantly the threat posed by Russell's discovery to his whole program of logical reduction.

Russell's interest in Frege's writings dates from this point. His correspondence with Frege and his study of his writings were essential for the reconstruction of his own logic on which he embarked in 1903. He therefore could write in *Principia Mathematica* that 'in all questions of logic our chief debt is to Frege' (vol. 1, p. viii). That was no exaggeration. The formal, axiomatic, truth-functional logic of *Principia* and its quantificational logic were deeply indebted to Frege. Russell had abandoned his earlier account of general propositions and adopted the Fregean, contextual analysis. The theory of definite descriptions which he first formulated in 1905—and which has often been considered the paradigm of analytic procedure—was in turn a further application of Frege's method of contextual analysis, a fact that Russell never adequately acknowledged. The theory of types which the authors of *Principia Mathematica* used to resolve the paradoxes had its root in another Fregean doctrine, that of the distinction of levels of function, for which there had been no equivalent in Russell's logical views of 1902.

When analytic philosophers trace the origins of their tradition back to Russell and Moore they often see in them a revival of British empiricism. There is no doubt that the analytic tradition began to incorporate a great deal of empiricism as it developed out of its beginnings, but the empiricist leanings were not so obvious at the outset either in Frege or in Russell and Moore. Only relatively late in Russell's development, with the publication of *Our Knowledge of the External World* in 1914, do the empiricist tendencies in his thought become evident.[15] And even then it is not clear to what extent it is correct to view Hume, for example, as the appropriate ancestor of the analytic conception of philosophy. Analytic philosophers themselves have typically cast Hume in that role, but Barry Stroud has argued in a recent study of Hume that his 'most direct legacy is not recent positivism or analytic empiricism . . . but rather that scientific naturalism that suffused so much of the thought about man, animals, and nature in the nineteenth century.'[16]

5 Carnap

The combination of the methodology of logical analysis with empiricist preconceptions is even more evident in Carnap and the other members of the Vienna Circle. Their logical empiricism was heavily influenced by Russell, but it had other equally strong empiricist sources from another direction.

The Vienna Circle originated in the empirically orientated tradition of

Austrian philosophy for which Mach was the principal spokesman. Through the influence of Carnap, who came to Vienna in 1926, and the reading of Wittgenstein's *Tractatus* there emerged the peculiar combination of logical and formal constructions with traditional empiricism that is now identified as the Vienna Circle.

Carnap had been one of Frege's students at Jena—in fact the only one who became known. He had also attended the lectures of Haeckel there. In the years after the First World War he studied the writings of Russell and had taken from them a deep conviction of the power of the new logic and its ability to reconstruct knowledge and science in a rational form. For Carnap, as for the whole Vienna Circle, the Russellian theory of types became an article of faith.

What Carnap owed to Frege, as he himself later declared,[17] was first of all an interest in logical syntax and semantics.

> Furthermore the following conception, which derives essentially from Frege, seemed to me of paramount importance. It is the task of logic and mathematics within the total system of knowledge to supply the forms of concepts, statements, and inferences, forms which are then applicable everywhere, hence also to non-logical knowledge. . . . This point of view is an important factor in the motivation of some of my philosophical positions, for example, . . . for my emphasis on the fundamental distinction between logical and non-logical knowledge.

Carnap credits Frege with this particular conception, but the separation of the formal and logical sub-structure from the empirical content of human knowledge and the belief that the formal element can be investigated *a priori* is, of course, ultimately due to Kant. It is clear from this statement that Carnap, and through him the whole tradition of logical positivism, has inherited at least part of Frege's rationalist program. In this respect it is surely correct to call Carnap 'Frege's legitimate successor.'[18]

In other respects, however, Carnap eventually developed less Kantian and less rationalist views than those of Frege. In his Ph.D. thesis, written under the direction of Bruno Bauch and Frege, Carnap maintained the Kantian (and Fregean) thesis that the propositions of geometry are synthetic *a priori*.[19] In the more empiricist-oriented climate of Vienna he came to abandon that view (*ibid.*, p. 50), while retaining the Fregean doctrine that 'all mathematical [i.e., presumably, arithmetical] concepts can be defined on the basis of the concepts of logic and that the theorems of mathematics can be deduced from the principles of logic' (*ibid.*, p. 46). But he gave that doctrine a conventionalist and formalist interpretation that was alien to the Fregean enterprise. He says that his view 'became more radical and precise, chiefly through the influence of Wittgenstein' (*ibid.*, p. 12).

Unlike Frege, Russell had originally had very little interest in language.

In the *Principles of Mathematics* (p. 47) he wrote: 'A proposition, unless it happens to be linguistic, does not itself contain words: it contains the entities indicated by words. Thus, meaning, in the sense in which words have meaning, is irrelevant to logic.' When he came to construct the formal symbolism of *Principia Mathematica* he thought of it more or less as Frege had thought of his own symbolism. For both the notation is a universal language with fixed meaning. Carnap's interest in language was similar in that it focused almost exclusively on formal, logical languages. In 1934 he wrote:[20]

> In consequence of the unsystematic and logically imperfect structure of the natural word-languages (such as German or Latin), the statement of their formal rules of formation and transformation would be so complicated that it would hardly be feasible in practice. . . . Owing to the deficiency of the word-languages, the logical syntax of language of this kind will not be developed, but, instead, we shall consider the syntax of two artificially constructed symbolic languages.

A similar doubt had been expressed slightly earlier by Tarski. When considering the clarification of the notion of truth, he had maintained that with respect to ordinary language 'not only does the definition of truth seem to be impossible, but even the consistent use of the concept in conformity with the laws of logic.'[21]

While Carnap's early work was exclusively concerned with the syntax of formal languages, Tarski's work set out to show that the semantics of such languages could also be described with logical rigor. The development of formal semantics involved a rejection of Frege's and Russell's conception of the symbolism as a language with fixed meaning and it also involved the rejection of the Fregean doctrine of the priority of judgments over concepts.

Löwenheim in his 1915 essay on 'The Possibilities in the Calculus of Relatives' was the first to revive the Boolean notion of the symbolism as a calculus capable of a variety of interpretations.[22] The development of model theory and possible world semantics derived from that new viewpoint lies outside the scope of Frege's ideas, although Dummett has argued that Frege's 'notion of reference coincides with the notion of interpretation . . . as currently employed in mathematical logic'[23] and has ascribed to Frege 'the modern distinction between the semantic (model-theoretic) and the syntactic (proof-theoretic) treatments of the notion of logical consequence' (*ibid.*, p. 80).

This judgment not only overlooks the difference between the Fregean views of logical symbolism and the Boolean conception incorporated in contemporary semantics, it also ignores the fact that semantics as it has developed from Tarski's work is strongly committed to philosophical assumptions that are antithetical to Frege's. Tarski's account of meaning is recursive, whereas Frege's is not. Tarski's account assumes a domain of objects as the fundamental given of the semantic theory and interprets all terms as denoting either objects of the domain or subsets of such objects or terms for relations between the non-logical objects. Tarski's philosophical views are indebted to the reism of Kotarbinski, which sees the world merely as an arrangement of objects. This philosophical viewpoint is, however, completely alien to Frege.

Dummett's interpretation of Frege as a precursor of modern semantics may be due in part to Carnap's account of Frege's doctrines. In his 1947 study of semantics and modal logic, *Meaning and Necessity*, Carnap explicitly referred to Tarski's work and declared himself 'in close agreement with his conception of the nature of semantics.'[24] It was also in this work that Carnap gave the first detailed account of Frege's semantic doctrines, treating them as a special version of the compositional theory of meaning that he was developing in accordance with Tarski's semantic methodology. In Carnap's account no reference is made to the Fregean dictum that words have meaning only in propositional contexts and it is not clear how a semantic theory of the Tarski type could be developed with that doctrine in mind.

Unlike Carnap, Dummett is aware of the significance of Frege's maxim but, since, like Carnap, his purpose is to present Frege's views in the context of the developments initiated by Tarski, he reduces the significance of that principle to merely expository importance, claiming that for Frege 'in the order of *explanation* the sense of the sentence is primary, but in the order of *recognition* the sense of a word is primary.'[25]

The history of Frege's dictum is instructive in showing how a doctrine that is at one moment of great significance in a philosophical tradition can be lost or change its meaning. When that has happened, no meaning analysis of the kind offered by analytic philosophers from Frege through Tarski and Carnap to Dummett can recover the actual historical meaning of the statement. That requires historical analysis of the kind provided in this study, an analysis that seeks to recover the original meaning by examining the historical origins of the statement, the role it plays in Frege's theorizing, and the significance attributed to its denial by Frege's contemporaries. Only then can the limits of the 'objective' unhistorical kind of meaning analysis conducted by analytic philosophers be overcome.

6 Wittgenstein

In the preface to the *Tractatus* Wittgenstein explicitly identifies two sources of his philosophizing—'the magnificent works of Frege and the writings of my friend Bertrand Russell.' The influence of Frege and Russell on the views of the *Tractatus* is in fact pervasive.

THE END OF LOGICISM

Like Frege and Russell, Wittgenstein is concerned with understanding the logical structure of language. 'The task of philosophy is the logical clarification of thought' (4.112). Philosophy so understood is not dependent on the natural sciences; its investigations are based neither on psychology (4.1121) nor on the theory of evolution (4.1122).

Wittgenstein believes with Frege that propositions have objective meaning. This objectivism shared by both of them has been interpreted as a form of realism. The first part of the *Tractatus* seems to lend itself easily to such a realist interpretation. Wittgenstein says there that the world has a certain logical structure and that language depicts that structure. Only the later sections of the book reveal the inadequacy of a realist interpretation of those claims. For Wittgenstein realism, idealism, and solipsism ultimately coincide. The world turns out to be my world and language my language. 'The limits of my language signify the limits of my world' (5.6). The objectivism which at first seemed straightforwardly realist must ultimately be understood in Kantian and transcendental terms.

Wittgenstein agrees with Frege that the philosophical clarification of thoughts will not issue in a theory. Both of them regard semantic considerations as purely practical and not theoretical. Philosophy proceeds through elucidations (4.112). We cannot define elementary terms, but can only give elucidations of them. And since the meaning of elementary signs is given only in the context of the sentences in which they occur, elucidations are nothing more than sentences in which elementary signs occur. 'Only a sentence has sense; only in the context of a sentence has a name meaning' (3.263).

Frege's doctrine that words have meaning only in propositional contexts deeply influenced both the early and the later thought of Wittgenstein. Of all the ideas he took from Frege that was the most significant for him. But in both his early and his late thought he used the doctrine in his own way. In the *Tractatus* he combined it with the thesis that sentences have an ultimate analysis. Frege had thought that every sentence can be analysed in more than one way and that there is no single correct analysis of a sentence. The logical structure we assign to a sentence is always relative to a set of other sentences. Wittgenstein, in contrast, holds that there are absolute simples and that sentences are built out of them in a definite way. Every sentence has an ultimate analysis. Its simple constituents are names and what they stand for are simple objects. Wittgenstein is clearly influenced by Russell in this atomistic doctrine, but his atomism differs from Russell's in one important respect. Russell's doctrine that simples are items of acquaintance is what ties his program of logical analysis to the traditional empiricist program of reducing all our ideas to combinations of sense data. For Wittgenstein simple objects are not at all items of acquaintance; they can only be postulated. Objects are those ultimate constituents which we must assume in order to understand the 'logical multiplicity' of our language.[26] His view is clearly more akin to Frege's than to Russell's.

Like Frege's notion of concept, Wittgenstein's notion is purely extensional. In this respect both differ from Russell, for whom propositional functions are intensional. In the first edition of *Principia Mathematica* Russell had adopted a ramified theory of types to resolve the logical contradictions. That theory depended on an intensional interpretation of concepts. In 1925 Ramsey proposed a modification and simplification of Russell's theory based on the thesis of extensionality which he had derived from the *Tractatus*.[27] The resulting simple theory of types, adopted by Russell in the second edition of *Principia Mathematica*, is substantially closer to Frege's theory of levels of functions than was the ramified theory. The thesis of extensionality also became a fundamental postulate of logical positivism.[28]

Even where Frege's and Wittgenstein's views diverge they often share common themes. Both speak of a distinction of sense and reference, but for Wittgenstein it is only sentences that have sense and names that have reference. Both are concerned with proper names and definite descriptions, but Wittgenstein follows Russell in regarding definite descriptions as incomplete symbols. Both consider truth-functionality essential to logic, but Wittgenstein goes further than Frege by trying to explain even universal and existential propositions in truth-functional terms. And both regard arithmetic as *a priori*, but Wittgenstein interprets arithmetical equations as pseudo-propositions.

While it is true that a multiplicity of ideas ties the *Tractatus* to the thought of Frege and Russell, it would be a mistake to assume that it is entirely derived from those two sources. Even though Wittgenstein was deeply influenced by Frege's and Russell's objectivism, logicism, and rationalism, he was also at the same time under the influence of Austrian naturalism. By the time he started to write the *Tractatus* he had probably read Mach and was certainly influenced by the views of Hertz, who in turn was indebted to Mach. Most important was the influence of Mauthner, whose *Beiträge zur Kritik der Sprache* he had read (cf. 4.0031).

With Mauthner Wittgenstein shared an admiration for Schopenhauer. What is often regarded as the influence of Schopenhauer on the later sections of the *Tractatus* may in fact be directly due to Mauthner. Wittgenstein's reflections on the self, the will, the limits of language, and the mystical seem closely connected to the themes Mauthner pursues in his *Beiträge*. The image of the ladder which one must eventually throw away at the end of the *Tractatus* is directly borrowed from Mauthner. And Mauthner ends the first volume of his *Beiträge* with a reflection that anticipates Wittgenstein's own conclusion: 'The critique of language must teach the liberation from language as the highest goal of the liberation of the self. Language thus becomes the self-criticism of philosophy.'[29]

There are in particular two important ideas which the early Wittgenstein shares with Mauthner. The first is that logical truths are empty tautologies. For Frege logical truths can extend our knowledge and different logical truths can express different thoughts. For Wittgenstein, as for Mauthner, logical truths are without sense (4.461). They are mere tautologies.[30] This characterization of logical truth was taken up by the logical positivists in their attempt to reconcile their empiricism with the rationalistic objectivism they had learnt from Frege and Russell. In the programmatic essay 'The Old and the New Logic' Carnap writes:[31]

> On the basis of the new logic one can clearly recognize the character of logical propositions. That is of the greatest significance for the epistemology of mathematics and for the clarification of many controversial philosophical questions. . . . It can be shown that all propositions of logic are tautologies and according to the view advocated here also all propositions of mathematics.

The other respect in which Wittgenstein is influenced by Mauthner is in his interest in natural language. For Frege, Russell, and Carnap the formal language of logic is of predominant interest. They conceive of a fundamental difference between the precise language of the logical symbolism and natural language. Mauthner, on the other hand, is a philosopher of ordinary language. In the period of the *Tractatus* Wittgenstein disagrees strongly with Mauthner's critique of formal logic. That is what he has in mind in writing: 'All philosophy is critique of language. But not in Mauthner's sense' (4.0031). But he agrees with Mauthner that a philosophical analysis of meaning must apply to all linguistic discourse and not only to logical symbolism. That view is combined with Frege's and Russell's reflections on logical structure in Wittgenstein's view that 'language disguises the thought, so that from the external form of the clothes one cannot infer the form of the thought they clothe, because the external form of the clothes is constructed with quite another object than to let the form of the body be recognized' (4.002). But underneath that external form ordinary language must have a perfect logical structure. 'All the sentences of ordinary language are actually logically completely in order, just as they are' (5.5563).

By the late 1920s Wittgenstein began to be increasingly worried by the rationalistic presuppositions of the *Tractatus* doctrine. The catalyst for the change in his views is likely to have been a lecture that the Dutch intuitionist Brouwer delivered at the University of Vienna in 1928.[32] In his lecture Brouwer sharply attacked both the trust in formal languages as superior to ordinary language and the idea that ordinary language was somehow logically in perfect order:

> There is no precision and certainty in the transmission of will, nor in

particular in the transmission of will through language. And this situation remains unchanged when one considers the transmission of will through the construction of purely mathematical systems. There is therefore no safe language for mathematics either. . . . The considerations of the formalist school have their origin in a superstitious faith in the magical power of language . . . a natural consequence of an older, more primary, more consequential and more deeply rooted error: namely, the careless trust in classical logic.

In Brouwer's view there is nothing ultimate about the laws of logic; their justification is merely practical.

As Wittgenstein's doubts about his earlier views increased, he discarded more and more of the assumptions he had inherited from Frege and Russell. His thinking drove him back to a view of language which was closely akin to those views of Mauthner's from which he had earlier distinguished the doctrines of the *Tractatus*.

Like Mauthner he came to be extremely critical of the attempt to understand natural language on the model of the formal calculus. In the *Philosophical Investigations* (§81) he writes:

If you say that our languages only *approximate* to such calculi you are standing on the very brink of a misunderstanding. For then it may look as if what we were talking about were an *ideal* language. As if our logic were, so to say, a logic for a vacuum.

Like Mauthner he seems at times unduly sceptical of the service that formal calculi can perform in the understanding of ordinary language and this scepticism has remained one of the controversial points in analytic philosophy of language. The historical significance of Wittgenstein's philosophy of language lies perhaps not so much in this scepticism itself but in the fact that it combines the two strands of the philosophy of language deriving from Mauthner and Frege. It is the interest in ordinary language combined with an interest in logical structure that is most characteristic of analytic philosophy since 1945.

Though Wittgenstein abandoned much of his Fregean heritage in his later philosophy he retained the idea that the meaning of the word is determined by the role it plays in language. He begins the *Blue Book*, the first expression of his later philosophy, with the question 'What is the meaning of a word?' The question turns out to have no direct and simple answer and is replaced by a series of new questions 'What is an explanation of meaning?', 'How is meaning learnt?', 'How are words used?', 'How is the word "meaning" used?' As we ask these questions Wittgenstein sets out to convince us that meaning is not an entity correlated with a word, nor is a word meaningful because it stands for something, but rather the meaning of a word is the use it has in language. Frege's context principle has been transformed into the claim that the study of language must itself be seen as part of the study of human practices.

Analytic philosophy arose in reaction to a dominant naturalism. From the very beginning it opposed radical empiricism, psychologism, historicism, evolutionism, and subjectivism. In contrast, it concerned itself with logical, formal, or *a priori* questions. As the tradition developed from Frege through Russell to Carnap and finally the later Wittgenstein it was forced to make greater concessions to the claims of empiricism. In Wittgenstein's later philosophy the tradition has reached a point at which it reconnects with the naturalism that emerged in the first half of the nineteenth century. Wittgenstein's philosophy of language finds its closest kin, not in the work of Frege or Carnap, but in the writings of Gruppe and Mauthner. It appears, then, that the development of analytic philosophy describes something like a circle. Frege thought he had banished radical empiricism, but in Wittgenstein it has returned to haunt the analytic tradition.

Wittgenstein's influence on contemporary analytic philosophy is pervasive and yet it seems that the analytic tradition has not so far drawn the most radical consequences out of Wittgenstein's thought. Instead it has tried to assimilate that thought to the logically oriented philosophy of language that originated with Frege. It has picked a multitude of insights out of Wittgenstein's philosophy without acknowledging that this philosophy undermines the view of the relation of logic, mathematics, and language that has prevailed in analytic philosophy since Frege. For Wittgenstein logic and mathematics are outgrowths of language and cannot be used to reveal the essence of language. According to Wittgenstein, that essence is revealed by neither a theory of objective meaning nor a theory of understanding. The essence of language shows itself only if we attend to the concrete uses of language. On this conception the abstract theory of meaning must give way in all but the most trivial cases to the examination of actual historical discourse.

The present study has been an attempt to show how analytic philosophers came to conceive of the analysis of meaning the way they do and how such an analysis might be carried out as an investigation of actual historical discourse.

Notes

Introduction

1 Cf. his essays on this topic, now collected in M. Dummett, *Truth and Other Enigmas*, Cambridge, Mass., 1978, and his monumental study, *Frege. The Philosophy of Language*, London, 1973. For a critical assessment of the historical verisimilitude of the latter, see H. D. Sluga, 'Frege and the Rise of Analytic Philosophy,' *Inquiry*, vol. 18, 1976, pp. 471–87.
2 M. Schlick, 'Der Wendepunkt der Philosophie,' *Erkenntnis*, vol. 1, 1930, pp. 4–11.
3 A. J. Ayer (ed.), *The Revolution in Philosophy*, London, 1956.
4 Dummett, 'Frege's Philosophy,' in *Truth and Other Enigmas*, p. 89; originally published as an article on Frege in P. Edwards (ed.), *Encyclopedia of Philosophy*, New York/London, 1967.
5 Dummett, *Frege. The Philosophy of Language*, p. 669.
6 Dummett, 'Frege as a Realist,' *Inquiry*, vol. 19, 1976, p. 468.
7 Dummett, *Frege. The Philosophy of Language*, p. xvii.
8 Cf. the essays on 'Frege's Ontology' in E. D. Klemke (ed.), *Essays on Frege*, Urbana/Chicago/London, 1968.
9 Dummett, *Frege. The Philosophy of Language*, p. 90; also p. 80.

Chapter I Philosophy in Question

1 M. Dummett, 'Frege, Gottlob,' in P. Edwards (ed.), *Encyclopedia of Philosophy*, New York/London, 1967, vol. 4, p. 225.
2 Dummett, *Frege. The Philosophy of Language*, London, 1973, p. 684.
3 For an examination of Dummett's arguments in support of his historical hypothesis, cf. H. D. Sluga, 'Frege's Alleged Realism,' *Inquiry*, vol. 20, 1977, pp. 227–42.
4 The recognition of the significance of the intellectual shift at the turn of the nineteenth century is, in the first instance, due to Michel Foucault, *The Order of Things*, New York, 1973.
5 'Animadversions on Descartes' Principles of Philosophy,' trans. G. M.

Duncan, in *The Philosophical Works of Leibniz*, 2nd edn, New Haven, 1908, p. 59.
6 G. W. F. Hegel, *Wissenschaft der Logik*, in *Werke*, vol. 3, Berlin, 1833, p. 62.
7 Frege's later claim that logic as a science of pure thought is fundamental to philosophy does not therefore signal a break with the philosophical tradition, but is a reaffirmation of it.
8 F. W. J. Schelling, *System des transzendentalen Idealismus*, in *Sämmtliche Werke*, vol. 3, pp. 444–50.
9 E. C. Agassiz, *Louis Agassiz: His Life and Correspondence*, Boston, 1885, pp. 151–2. Quoted from S. J. Gould, *Ontogeny and Phylogeny*, Cambridge, Mass./London, 1977, p. 39.
10 The generational conflict coming out of these changes has been described very well in Königsberger's biography of Hermann von Helmholtz: 'The more the young man's thoughts, the direction of his labours, and his whole scientific attitude (which was soon to be adopted by the entire world of science) took him away from metaphysical speculation [around the time of 1843], the stronger and for some time irreconcilable became the contrast with the wholly speculative philosophy of his father. While Ferdinand Helmholtz admitted only the deductive method in science, and held inductive reason inimical to it, Hermann on the contrary bore the latter upon his shield, proclaiming it to the end of his life the salvation of science in general and not merely of the physical sciences. The father (secure in the consciousness that he must be the better able as a philosopher to appreciate the relation in which man stands to experience, and with the best intention in the world of directing his son, his "dearest treasure," into the right paths of scientific discovery) missed no opportunity in their daily intercourse of bringing his general philosophical convictions and metaphysical conceptions to bear upon the young man, doing all he could to shake him in his methods of thought and experiment' (L. Königsberger, *Hermann von Helmholtz*, New York, 1965, p. 30).
11 Cf. F. M. Turner, *Between Science and Religion. The Reaction to Scientific Naturalism in Late Victorian England*, New Haven/London, 1974.
12 J. von Liebig, *Über das Studium der Naturwissenschaften und über den Zustand der Chemie in Preussen*, Braunschweig, 1884, p. 44. Quoted from F. Lilge, *The Abuse of Learning. The Failure of the German University*, New York, 1948, pp. 60–1.
13 R. Haym, *Hegel und seine Zeit*, 2nd edn Leipzig, 1927, p. 4. First published 1857.
14 M. J. Schleiden, *Über den Materialismus der neueren Naturwissenschaft, sein Wesen und seine Geschichte*, Leipzig, 1863, p. 57.
15 J. Moleschott, *Ursache und Wirkung in der Lehre vom Leben*, Giessen, 1867, p. 12. Quoted from F. Gregory, *Scientific Materialism in Nineteenth Century Germany*, Dordrecht/Boston, 1977, p. 146.
16 W. Wundt, *Erlebtes und Erkanntes*, Leipzig, 1920, pp. 238–42.
17 L. Feuerbach, *The Essence of Christianity*, preface to 2nd edn.
18 Moleschott's book *Die Lehre der Nahrungsmittel* appeared in 1850 and was enthusiastically reviewed by Feuerbach. In 1855 Büchner published the most widely read work of this new naturalism, *Kraft und Stoff*. In the same year Vogt presented his side of the controversy about scientific naturalism that had

erupted at the 1853 congress of scientists at Göttingen. His book was called *Köhlerglaube und Wissenschaft*. Czolbe's *Neue Darstellung des Sensualismus* also appeared in 1855. It was the most philosophical, but also the most idiosyncratic, exposition of the new thought.
19 Cf. R. W. Göldel, *Die Lehre von der Identität*, Leipzig, 1935, pp. 21–2.
20 L. Büchner, *Aus Natur und Wissenschaft*, vol. 2, Leipzig, 1884, p. 252.
21 Cf. Wundt, *op. cit.*, pp. 222–3.
22 For a wealth of material, cf. F. Gregory's thorough and enlightening study *Scientific Materialism in Nineteenth Century Germany*, Dordrecht/Boston, 1977.
23 J. G. Herder, *Verstand und Erfahrung. Eine Metakritik zur Kritik der reinen Vernunft*, 1799. Quoted from J. G. Herder, *Sprachphilosophische Schriften*, ed. E. Heintel, Hamburg, 1953, p. 183.
24 J. G. Herder, *Abhandlung über den Ursprung der Sprache*, quoted from Heintel, p. 50.
25 Republished in 1914 with an instructive introduction by Fritz Mauthner, who recognized in Gruppe an important forerunner of his own philosophy of language.
26 In 1857 Büchner reviewed Gruppe's third book, giving it the highest praise for its attack on speculative philosophy and completely agreeing with its rejection of *a priori* reasoning. Cf. L. Büchner, *Aus Natur und Wissenschaft*, Leipzig, 1874, pp. 37–41.
27 J. S. Mill, *An Examination of Sir William Hamilton's Philosophy*, New York, 1973, vol. 2, ch. 20, pp. 245–6.
28 F. A. Lange, *Geschichte des Materialismus*, Frankfurt, 1974, vol. 2, p. 552.
29 H. Czolbe, *Neue Darstellung des Sensualismus*, Leipzig, 1855, pp. 233–4.
30 Lange, *loc. cit.*
31 Wundt, *op. cit.*, p. 129.
32 *Ibid.*
33 H. Lotze, Recension von Heinrich Czolbe, 'Neue Darstellung des Sensualismus,' in *Kleine Schriften*, vol. 3, Leipzig, 1891, p. 243.
34 H. Lotze, Recension von Heinrich Czolbe, 'Entstehung des Selbstbewusstseins. Eine Antwort an Herrn Professor Lotze,' in *Kleine Schriften*, vol. 3, p. 316.
35 H. Vaihinger, 'Die drei Phasen des Czolbeschen Naturalismus,' *Philosophische Monatshefte*, vol. 12, 1876, pp. 1–31.
36 Frege elaborates the same idea again in his late essay 'The Thought,' KS, pp. 355–7.
37 Lotze, Recension von Heinrich Czolbe, 'Neue Darstellung des Sensualismus,' p. 240.
38 *Ibid.*, p. 241.
39 Cf. the account given in Harald Höffding's *History of Modern Philosophy*, vol. 3, London, 1900, pp. 508–24.
40 Most unexpected anomalies of this kind occur. I remember from my student days at Munich around 1960 a circle of what can only be called *believing Fichteans*, who could occasionally be heard at philosophical gatherings.
41 One can read Descartes and label his time 'The Age of Reason'; one can also look at Callot's contemporary etchings of the horrors of war or read Huxley's *Devils of Loudun* and form quite a different picture of the same period.

NOTES TO PAGES 33-37

42 Kuhn's description of the history of science as a sequence of discontinuous periods separated by incommensurable paradigms lends itself to this perverse form of historiography. Even more so do the writings of the French structuralists with their endless search for hidden deep structures. Thus, Foucault (in *The Order of Things*), having decided that in the break from the eighteenth to the nineteenth century one unified deep structure (or '*episteme*') has replaced another (p. 217), finds himself in some embarrassment to define the *episteme* of the new period. He finally identifies it as 'the criticism-positivism-metaphysics triangle of the object' (p. 245) and again as 'this double affirmation—alternating or simultaneous—of being able and not being able to formalize the empirical' (p. 246). The emptiness of such formulas shows adequately how useless it is to search for Hegelian ground plans in history, particularly for periods as complex and diverse as the nineteenth century.
43 From America there had begun to spread a wave of interest in spiritualism and psychical research, which were often taken to disprove the claims of the materialists and naturalists. Cf. Turner, *op. cit.*, pp. 50–60, 84–103, 117–33.
44 Cf. H. Stuart Hughes, *Consciousness and Society. The Reorientation of European Social Thought 1875–1930*, New York, 1958.
45 Cf. F. Paulsen, 'Ernst Haeckel als Philosoph,' *Preussische Jahrbücher*, vol. 101, 1900, pp. 29–72.

Chapter II   Philosophical Reconstruction

1 'Über die Gründe der Entmutigung auf philosophischem Gebiete,' reprinted in F. Brentano, *Über die Zukunft der Philosophie*, ed. O. Kraus, Leipzig, 1929.
2 The phrase was coined by Otto Liebmann in his book *Kant und die Epigonen*, which first appeared in 1865.
3 There are other important figures of post-naturalist thought who are left out of the present discussion. Nietzsche and Dilthey come readily to mind.
4 A clear exposition of such a critique of materialism can be found in F. A. Lange's admirable *History of Materialism*, which first appeared in 1866 and in a second, considerably revised, version in 1872.
5 This revival of the philosophical tradition did not, characteristically, take the form of a straightforward return to the past. The prevailing scientific climate forced the Neo-Kantian philosophers into making suitable adjustments to the original Kantian doctrines. For the same reason, there was no reawakening of Hegelianism in Germany at that time.
6 Cf. B. Bauch, *Jena und die Philosophie des deutschen Idealismus*, Jena, 1922.
7 Thus, characteristically, the journal *Annalen der Philosophie*, which had been published as the organ of Hans Vaihinger's philosophy of the as-if and which through him had loosely Kantian affiliations, became in 1930 the official publication of the logical positivists under its new name *Erkenntnis*.
8 On the 'powerful and ineffaceable impression' that Kant made on Mach, see E. Mach, *The Analysis of Sensations* (Dover edn), New York, 1959, p. 30, fn.
9 E. Mach, *The Science of Mechanics*, La Salle, Ill., 1960, p. 579.
10 E. Mach, *Space and Geometry*, La Salle, Ill., 1960, p. 98.

NOTES TO PAGES 38-49

11 Cf. W. Jerusalem, *Introduction to Philosophy*, New York, 1910, p. 51.
12 E. Husserl, *Logische Untersuchungen*, Tübingen, 1968, vol. 1, p. 203.
13 W. Jerusalem, *Der kritische Idealismus und die reine Logik*, Vienna/Leipzig, 1905, p. 95.
14 Husserl, *Logische Untersuchungen*, 'Prolegomena zur reinen Logik.'
15 P. Frank, *Between Physics and Philosophy*, Harvard, 1941, p. 187.
16 J. T. Blackmore, *Ernst Mach*, Los Angeles/London, 1972, p. 186.
17 Cf. F. Nietzsche, *Beyond Good and Evil*, §20.
18 For an account of the origins of Mauthner's thought, cf. F. Mauthner, *Prager Jugendjahre*, Frankfurt, 1969, pp. 200-22.
19 E. Mach, *Erkenntnis und Irrtum*, Leipzig, 1905; W. Jerusalem, *Der kritische Idealismus*, p. 180.
20 For details of Mauthner's philosophy of language, cf. the following chapter.
21 Cf. Husserl, *Logische Untersuchungen*, vol. 1, p. xii.
22 P. F. Linke, 'Das Recht der Phänomenologie,' *Kantstudien*, vol. 21, 1916, p. 207.
23 Husserl, *Logische Untersuchungen*, vol. 1, p. 212.
24 Ernst Abbe, Frege's mentor, became famous for his work in mathematical optics, which had both theoretical interest for the theory of light and practical significance for the design of better lenses.
25 For details on these (philosophically and mathematically) significant developments see O. Becker, *Die Grundlagen der Mathematik in geschichtlicher Entwicklung*, Freiburg/München, 1954, pp. 168-316.
26 One must read these critical remarks against the background of Cantor's completely unsympathetic review of Frege's *Grundlagen* (published originally in 1885 and reprinted in G. Cantor, *Gesammelte Abhandlungen*, Hildesheim, 1962, pp. 440-1).
27 To gain an overview of Frege's interests it is not sufficient to study the works he published or those that have been published posthumously. Since much of the unpublished material was destroyed in the Second World War, one must also consult the annotated list of manuscripts compiled by Heinrich Scholz sometime before the war. It was published under the title 'Verzeichnis des wissenschaftlichen Nachlasses von Gottlob Frege,' in M. Schirn (ed.), *Studies on Frege*, vol. 1, Stuttgart/Bad Canstatt, 1976, pp. 86-103.
28 *Ibid.*, p. 100.
29 Becker, *op. cit.*, pp. 178-83.
30 *Ibid.*, p. 183.
31 *Ibid.*, p. 179.
32 Scholz, *op. cit.*, p. 90.
33 T. W. Bynum, 'On the Life and Work of Gottlob Frege,' in G. Frege, *Conceptual Notation*, ed. T. W. Bynum, Oxford, 1972, p. 5.
34 Apart from some similarities in content between Boolean algebra and the logic of the *Begriffsschrift* (which will be considered further in the next chapter), the acquaintance with Boolean algebra is also suggested by the following comment about the *Begriffsschrift* from 1880/81: 'In a small pamphlet I have tried to get close again to the Leibnizian idea of a *lingua characterica*. In it I had to deal with some of the same topics as Boole, though in a different manner' (NS, pp. 11-12). Cf. also BS, p. x.

NOTES TO PAGES 49-60

35 A. Trendelenburg, Preface to the second edition, *Logische Untersuchungen*, 3rd edn, Leipzig, 1870, vol. 1, p. ix.
36 *Ibid.*, vol. 2, pp. 228-38.
37 Cf. BS, p. xi. Trendelenburg's essay appeared in *Historische Beiträge zur Philosophie*, vol. 3, Berlin, 1867.
38 In the essay 'Booles rechnende Logik und die Begriffsschrift' of 1880/81 Frege speaks of Leibniz's *lingua characterica* (cf. note 34 above), a term which Trendelenburg had used in his essay, instead of Leibniz's own term *lingua characteristica*. G. Patzig has suggested that Frege took this usage from Trendelenburg, though it may also have come to him from the older Leibniz editions of Raspe and Erdmann. Cf. NS, p. 9, note 2.
39 Heinrich Scholz was the first to notice this point; cf. his note in BS, p. 115.
40 Trendelenburg, *Historische Beiträge*, p. 1.
41 G. W. Leibniz, *Selections*, ed. P. P. Wiener, New York, 1951, p. 8.
42 A. Trendelenburg, *Historische Beiträge*, p. 3.
43 In our time J. L. Borges has taken pleasure in the 'rational mysticism' of these authors. Cf., e.g., his short and entertaining essay 'The Analytical Language of John Wilkins,' in J. L. Borges, *Other Inquisitions. 1937-1952*, New York, 1966, pp. 106-10.
44 G. Frege, 'Über die wissenschaftliche Berechtigung einer Begriffsschrift,' reprinted in BS, pp. 113-14.
45 The theory of judgment that he examines at the beginning of that book is in fact the Kantian theory in the particular form that Lotze had given it. This observation is due to Marcus Bierich, *Freges Lehre von dem Sinn und der Bedeutung der Urteile und Russells Kritik an dieser Lehre*, dissertation, Hamburg, 1951.
46 In R. Schmidt (ed.), *Die Philosophie der Gegenwart in Selbstdarstellungen*, vol. 7, Leipzig, 1929, p. 29.
47 B. Bauch, 'Lotzes Logik und ihre Bedeutung im deutschen Idealismus,' *Beiträge zur Philosophie des Deutschen Idealismus*, vol. 1, 1918, p. 45. The essay, in which Bauch explicitly refers to Frege, immediately precedes Frege's essay 'Der Gedanke' ['The Thought'], in the same issue of *Beiträge*. Bauch's discussion is obviously meant by the editors of the journal as a kind of introduction to Frege's piece.
48 M. Heidegger, 'Neuere Forschungen über Logik,' *Literarische Rundschau für das katholische Deutschland*, vol. 38, 1912, col. 469, footnote 2.
49 I. Kant, *Critique of Pure Reason*, A 54, B 78.
50 Lotze, *Logik*, 1880, pp. 11-12.
51 Cf. W. Dilthey, 'Erfahren und Denken. Eine Studie zur erkenntnistheoretischen Logik des neunzehnten Jahrhunderts,' *Gesammelte Schriften*, vol. 5, Stuttgart/Göttingen, 1957, pp. 74-89.
52 W. Wundt, *Erlebtes und Erkanntes*, Leipzig, 1920, p. 265.
53 Bauch, *op. cit.*, p. 47.
54 *Beiträge zur Philosophie des deutschen Idealismus*, vol. 1, no. 1, back cover.
55 Behind the idealistic sympathies of the society and its organ was a decidedly nationalistic temperament. Their publication was designed to give special emphasis to 'the national values of the German philosophical spiritual life' (*ibid.*). And they explicitly rejected the 'supra-nationalism' of other German

publications in philosophy. From the few things one knows about Frege's (later?) political views, one fears that it was not only the idealism that attached him to the society.

56 P. Natorp, *Die Grundlagen der exakten Wissenschaft*, 3rd edn, Leipzig/Berlin, 1923.
57 I. Kant, *Groundwork to the Metaphysics of Morals*, pp. 387–8.
58 L. Wittgenstein, *Philosophical Investigations*, §97.

Chapter III   A Language of Pure Thought

1 W. and M. Kneale, *The Development of Logic*, Oxford, 1962, p. 436.
2 I. M. Bochenski, *Formale Logik*, Freiburg/Munich, 1956, pp. 313–14.
3 G. Frege, *Conceptual Notation*, ed. T. W. Bynum, Oxford, 1972, p. 212.
4 T. W. Bynum, 'On the Life and Work of Gottlob Frege,' in *Conceptual Notation*, p. 16.
5 Frege, *Conceptual Notation*, p. 218.
6 H. Lotze, *Logik*, 2nd edn, Leipzig, 1880, p. 256.
7 Lotze, *Logik*, Leipzig, 1843, p. 49.
8 Lotze, *Logik*, 1880, p. 20.
9 F. Mauthner, *Wörterbuch der Philosophie*, 2nd edn, vol. 1, Leipzig, 1923, pp. 24–5.
10 Mauthner, *Die Sprache*, Frankfurt, 1906, pp. 30–1.
11 Mauthner, *Wörterbuch der Philosophie*, op. cit., p. xxxii.
12 Mauthner, *Beiträge zu einer Kritik der Sprache*, vol. 1, Stuttgart, 1901, p. 26.
13 In 1901 Couturat wrote to Frege about the adoption of an international auxiliary language. Such a language, he explained, would fill an urgent need. 'Our enterprise has already received . . . the support of very important societies, such as the Touring Club' (WB, p. 22). It would be important for both commerce and science. For that reason the scientific community had responded quite favorably to the promotion of the idea. Frege seems to have been cautiously supportive of the project while voicing some doubts about how new words were to be invented (*ibid.*, p. 23). The project of such an international language remained part of the logical positivist creed for more that thirty years.
14 Characteristically, it was Ostwald who was willing to publish Wittgenstein's *Tractatus* in his *Annalen der Naturphilosophie* after several publishers had turned it down. If Ostwald found Wittgenstein's ideas to his liking, the sentiment was not reciprocated.
15 W. Jerusalem, *Der kritische Idealismus und die reine Logik*, Vienna/Leipzig, 1905, p. 180.
16 K. Gödel, 'Russell's Mathematical Logic,' in P. Schilpp (ed.), *The Philosophy of Bertrand Russell*, New York, 1944, p. 126.
17 Cf. J. Lukasiewicz, 'On the History of the Logic of Propositions,' S. McCall (ed.), *Polish Logic 1920–1939*, Oxford, 1967, pp. 84ff.
18 The discussion of the notion of sameness of content (or identity) which Frege introduces into the *Begriffsschrift* will be taken up in the context of the examination of his later doctrine of sense and reference (see chapter V).
19 Leibniz, *Nouveaux Essais*, book iv, ch. viii, §10.

NOTES TO PAGES 83–108

20 The medieval logicians tried to resolve these difficulties by means of their theory of supposition. In his *Principles of Mathematics* of 1903 Russell was still struggling to rescue the traditional account by means of his theory of denoting. Peter Geach in *Reference and Generality*, Ithaca/New York, 1962, has given a detailed critical account of these efforts.
21 Quoted from O. Becker, *Die Grundlagen der Mathematik in geschichtlicher Entwicklung*, Freiburg/München, 1954, p. 219.
22 Cf. C. Thiel, *Sinn und Bedeutung in der Logik Gottlob Freges*, Meisenheim, 1965, pp. 17–18.
23 Several interpreters have stressed the similarity between Kant's and Frege's views. Cf. NS, p. 75, also M. Dummett, *Frege. The Philosophy of Language*, London, 1973, p. 278.
24 For Frege cf. R. Carnap, 'Intellectual Autobiography,' in P. Schilpp (ed.), *The Philosophy of Rudolf Carnap*, La Salle, Ill./London, 1963.
25 For a discussion of the various positions, cf. Thiel, *Sinn und Bedeutung*, pp. 124ff.
26 I. Hacking, *Why Does Language Matter to Philosophy?*, Cambridge, 1975, p. 1.

Chapter IV   In Search of Logical Objects

1 The Fregean notion of abstraction corresponds to the Kantian one. Cf. I. Kant, *Logik*, in *Gesammelte Schriften*, vol. 9, Berlin, 1923, pp. 94–5.
2 In dealing with them Frege for the first time proves himself an accomplished polemicist. It is a talent of which he made (perhaps too) frequent use in his later writings. His technique is generally the same: he takes remarks by his opponents absolutely literally; he draws consequences with absolute rigor; he brings large-scale claims down to earth with very concrete counter-examples; he searches out hidden inconsistencies. These techniques are often successful, sometimes devastating, and almost always rhetorically effective. But they do not usually lend themselves to a sympathetic assessment of divergent views.
3 A representative sample of essays on the topic of Frege's ontology can be found in E. D. Klemke (ed.), *Essays on Frege*, Urbana/Chicago/London, 1968. For a critical assessment, cf. my review in *Philosophy*, vol. 45, 1970, p. 75.
4 M. Dummett, 'Frege as a Realist,' *Inquiry*, vol. 19, 1976, p. 455.
5 Dummett, 'Platonism,' in *Truth and Other Enigmas*, Cambridge, Mass., 1978, p. 202.
6 Dummett, 'Realism,' *ibid.*, p. 146.
7 Dummett, Preface, *ibid.*, p. xxvii.
8 Cf. H. D. Sluga, 'Frege and the Rise of Analytic Philosophy,' *Inquiry*, vol. 18, 1975, p. 477.
9 Cf. the writings of the Dutch intuitionists van Dantzig and Griss. Their views are discussed in A. Fraenkel and Y. Bar Hillel, *Foundations of Set Theory*, Amsterdam, 1958, pp. 239–44.
10 Austin translates Frege's term *'allgemein'* throughout the *Foundations* as 'general' rather than as 'universal.' This translation fails to capture the full meaning of the term for Frege. In particular it fails to bring out the relation between Frege's claim that truths are *a priori* in so far as they are derivable

from universal laws and Kant's use of strict universality as one of the criteria (indeed as *the* criterion) of *a priori* truth (cf. *Critique of Pure Reason*, B 3–4).

11 Cf. his arguments in B. Russell, *Introduction to Mathematical Philosophy*, London, 1919, pp. 137ff.

12 F. P. Ramsey, 'The Foundations of Mathematics,' in *The Foundations of Mathematics and Other Logical Essays*, London, 1931, pp. 59–61.

13 Unfortunately the fragment cannot be dated with absolute certainty. It was written after the completion of the *Begriffsschrift* but before the development of the sense-reference semantics. The editors of the *Nachgelassene Schriften* therefore simply date it as between 1879 and 1891 (NS, p. 1). But there are several reasons for thinking that it was not written immediately after the *Begriffsschrift*. One is the terminological switch described above, which was certainly made after 1880/81. Another is that one of the projected sections of the text was supposed to deal with the definition of objects through judgments of re-identification (*ibid.*). That topic has no equivalent in the *Begriffsschrift*, but is closely connected to issues discussed in *Foundations*. My conclusion is that the stress on objective truth and on the problem of the definition of objects makes it likely that the text belongs to the period of 1884.

14 A fragment of a fourth essay in this series, on the topic of logical universality, was published posthumously (NS, pp. 278–81).

15 It is occasionally thought that the ideas expressed in Frege's late essays of 1918 to 1923 represent a break with some of his earlier preconceptions. As against that it is important to keep in mind that the late essays contain ideas and formulations that are already contained in the earlier projects. In these writings Frege, as always, reveals a surprising consistency in his views. The similarities are particularly close between the 1897 project and the essay 'The Thought'; so close, in fact, that we must consider the earlier text to be the draft from which Frege composed the essay.

16 Cf. J. v. Heijenoort, 'Logic as Calculus and Logic as Language,' *Synthese*, vol. 17, 1967.

17 Frege's term '*wirklich*' derives from '*wirken*' — to bring about. What is '*wirklich*' enters chains of causal interaction and is (therefore?) spatio-temporal (KS, pp. 360–1). Does this mean that '*wirklich*' is to be translated as 'actual'? Or could it be translated as 'real'? Dummett has called the latter a 'tendentious translation' ('Frege as a Realist,' p. 457). But is it mistaken? The problem is that '*wirklich*' is a perfectly legitimate translation for either 'actual' or 'real.'

18 Austin's usually admirable translation seems to miss out some of the meaning of this passage.

19 Cf. C. Thiel, *Sense and Reference in Frege's Logic*, Dordrecht, 1968, ch. viii.

20 Greg Currie has pointed out that in 'The Thought' Frege says nothing about the objectivity of thoughts. Instead he says that they must be *wirklich* in a certain way (cf. G. Currie, 'Frege's Realism,' *Inquiry*, vol. 21, 1978, pp. 218–21). At first sight that remark might seem to be incompatible with the interpretation I have offered. However, on closer inspection it turns out that Frege is not saying that thoughts themselves are real, but they are real only in so far as they are grasped or can be grasped. It is our ideas of thoughts that are

strictly speaking real and through which thoughts causally affect reality.
21 In the final pages of the *Foundations* Frege indicates that other types of number should be defined in an analogous manner (F, p. 114). Real and complex numbers are also to be characterized as extensions of concepts. In the second volume of the *Grundgesetze* Frege shows how this program can be carried through for the real numbers. He argues that we must first consider their use. Real numbers, he concludes from this consideration, are primarily used for measuring. And to measure is to establish a ratio between something and a measure. The real numbers can therefore be defined as ratios of magnitudes and that means as relations of relations (GG, vol. 2, p. 160). Since relations are double value-ranges in Frege's terminology, we have then defined the real numbers as value-ranges.

The general interest of this definition of the real numbers is that Frege achieves it by applying the same methodology he had used in his account of natural numbers. That methodology involved the principle that words have meaning only in propositional contexts. It seems then that Frege was still making practical use of the context principle in 1903.

Chapter V The Analysis of Meaning

1 A. Trendelenburg, *Logische Untersuchungen*, vol. 2, 3rd edn, Leipzig, 1870, p. 237.
2 When Frege describes the distinction between concept and object in the essay 'On Concept and Object' and calls it logically simple he repeats almost verbatim what he had said about the logically simple in 'On the Principle of Inertia' (cf. KS, pp. 124, 167).
3 M. Dummett, *Frege. The Philosophy of Language*, London, 1973, p. 628. Max Black had previously made much the same point (M. Black, *A Companion to Wittgenstein's* Tractatus, Cambridge, 1964, p. 117).
4 The essay has received some attention from philosophers of science, cf. P. Janich, 'Trägheitsgesetz und Inertialsystem—Zur Kritik G. Freges an der Definition L. Langes,' in M. Schirn (ed.), *Studies on Frege*, vol. 3, Stuttgart/Bad Canstatt, 1976, pp. 145–56. Further references are given there. Frege's essay has not received any attention so far in the context in which I discuss it here.
5 Cf. Wittgenstein, *Philosophical Investigations*, § 20: 'But now it looks as if when someone says "Bring me a slab" he could mean this expression as *one* long word.... I think we shall be inclined to say: we mean the sentence as *four* words when we use it in contrast with other sentences such as "*Hand* me a slab," "Bring *him* a slab," "Bring *two* slabs," etc.... We say that we use the command in contrast with other sentences because *our language* contains the possibility of those other sentences.'
6 This resolution of the difficulty was repeatedly urged by P. T. Geach before the *Nachlass* was published. Cf., e.g., P. T. Geach, 'Frege', in G. E. M. Anscombe and P. T. Geach, *Three Philosophers*, Oxford, 1961, p. 156.
7 Cf. K. Gödel, 'Russell's Mathematical Logic,' in P. Schilpp (ed.), *The Philosophy of Bertrand Russell*, New York, 1951, p. 149.
8 The affinity between Frege's and Wittgenstein's views on this point has been noted by Geach, 'Frege' in Anscombe–Geach, *Three Philosophers*, p. 147.

NOTES TO PAGES 145–167

9 Dummett, *op. cit.*, pp. 644–5.
10 G. Cantor, 'Beiträge zur Begründung der transfiniten Mengenlehre. I,' *Mathematische Annalen*, vol. 46, 1895, p. 481.
11 G. Cantor, 'Über unendliche lineare Punktmannigfaltigkeiten. V,' *Mathematische Annalen*, vol. 21, 1883, p. 587.
12 E. Schröder, *Vorlesungen über die Algebra der Logik*, vol. 1, Leipzig, 1890, p. 188.
13 A. Church, 'Schröder's Anticipation of a Simple Theory of Types,' *Journal of Unified Science*, vol. 9, 1939.
14 G. W. Leibniz, *Selections*, ed. P. P. Wiener, New York, 1951, p. 10.
15 Frege nowhere explains why the mode of determination is 'contained' in the sense rather than being identical with it. It therefore seems best to ignore the suggested distinction.
16 For Frege's equation of content and reference, see KS, p. 126.
17 Dummett, *op. cit.*, p. xvi.
18 E. Tugendhat, 'The Meaning of "*Bedeutung*" in Frege,' *Analysis*, vol. 30, 1970, p. 177.
19 In a similar spirit I. Angelelli has proposed the translation 'importance,' cf. *Studies on Gottlob Frege and Traditional Philosophy*, Dordrecht, 1967, p. 55.
20 Even when Frege expresses himself in terms that seem at first sight barely compatible with the thesis that sentence meaning is primary, the justification for assigning sense to names turns out to be that they must have a sense because they must make a contribution to the sense expressed by the whole sentence (cf. WB, p. 127).
21 Black and Geach have speculated that this further discussion is to be found in the essay 'On Concept and Object'; cf. M. Black and P. T. Geach, *Translations from the Philosophical Writings of Gottlob Frege*, Oxford, 1977, p. 57; but there is no discussion in that essay of the distinction of sense and reference for functional expressions. It seems more plausible to assume that Frege was referring to a projected sequel to the essay 'On Sense and Reference' which remained unpublished. Notes for this sequel have survived and are published in NS, pp. 128–36.
22 For an examination of the reasons that make the de-epistemologizing of semantics plausible and for a critique of Kripke's theory, cf. L. Foy, *Reference and Possible Worlds*, Ph.D. dissertation, Cornell University, 1978.

Chapter VI   The End of Logicism

1 B. Russell, *The Principles of Mathematics*, 2nd edn, London, 1956, p. 528.
2 Frege's remark means that concepts are extensional, i.e., that they coincide when their extensions are identical and that the expression 'the concept F' denotes an object and not like 'x is an F' a concept. The note suggests that numbers might be defined as objects denoted by expressions of the form 'the concept F' rather than as extensions of concepts.
3 H. Scholz, 'Verzeichnis des wissenschaftlichen Nachlasses von Gottlob Frege,' in M. Schirn (ed.), *Studies on Frege*, vol. 1, Stuttgart/Bad Canstatt, 1975, p. 96.
4 The flaw in Frege's argument was first exposed by J. Bartlett, *Funktion und Gegenstand*, Ph.D. dissertation, Munich, 1961, pp. 71ff.

NOTES TO PAGES 168-184

5 J. van Heijenoort, *From Frege to Gödel*, Cambridge, Mass., 1967, p. 127.
6 Cf. W. V. Quine, 'On Frege's Way Out,' *Selected Logical Papers*, New York, 1966, pp. 151-2.
7 R. Carnap, 'Intellectual Autobiography,' in P. Schilpp (ed.), *The Philosophy of Rudolf Carnap*, La Salle, Ill./London, 1963, p. 5.
8 Cf. C. Parsons, 'Some Remarks on Frege's Conception of Extension,' in M. Schirn (ed.), *Studies on Frege*, vol. 1, Stuttgart/Bad Canstatt, 1976, pp. 274-5.
9 B. Russell, *My Philosophical Development*, London, 1959, p. 38.
10 B. Russell, 'My Mental Development,' in P. Schilpp (ed.), *The Philosophy of Bertrand Russell*, New York, 1944, p. 13.
11 Russell, *My Philosophical Development*, p. 62.
12 G. E. Moore, 'The Nature of Judgment,' *Mind*, 1899, p. 192.
13 Cf. P. T. Geach, *Reference and Generality*, Ithaca, N.Y., 1962, for a good critical account of the theory.
14 Cf. P. Nidditch, 'Peano and the Recognition of Frege,' *Mind*, vol. 72, 1963, pp. 103-10.
15 Cf. P. Hylton's instructive Ph.D. thesis on *The Origins of Analytic Philosophy*, Harvard, 1978.
16 B. Stroud, *Hume*, London/Henley/Boston, 1977, p. 222.
17 Carnap, 'Intellectual Autobiography,' pp. 12-13.
18 E. W. Beth, 'Carnap on Constructed Systems,' in Schilpp, *op. cit.*, p. 472.
19 Carnap, 'Intellectual Autobiography,' p. 12.
20 R. Carnap, *The Logical Syntax of Language*, London, 1959, pp. 2-3.
21 A. Tarski, 'The Concept of Truth in Formalized Languages,' in *Logic, Semantics, Metamathematics*, Oxford, 1956, p. 153.
22 Cf. J. van Heijenoort, 'Logic as Calculus and Logic as Language,' *Synthèse*, vol. 17, 1967.
23 M. Dummett, *Frege. The Philosophy of Language*, London, 1973, p. 90.
24 R. Carnap, *Meaning and Necessity*, Chicago, 2nd enlarged edn, 1960, p. 64.
25 Dummett, *Frege. The Philosophy of Language*, p. 4.
26 Cf. F. Waismann, *Wittgenstein und der Wiener Kreis*, Oxford, 1967, pp. 41-3.
27 Cf. F. P. Ramsey, 'The Foundations of Mathematics,' in *The Foundations of Mathematics and Other Logical Essays*, London, 1954, pp. 49ff.
28 Cf. R. Carnap, *Der Logische Aufbau der Welt*, 2nd edn, Hamburg, 1961, §43.
29 F. Mauthner, *Beiträge zur Kritik der Sprache*, vol. 1, 3rd edn, Leipzig, 1923, p. 713.
30 Mauthner first made that claim in the third volume of the *Beiträge*, which originally appeared in 1913 (cf. 3rd edn, p. 321). In the later *Wörterbuch der Philosophie* he gives the idea even more prominence by beginning the first volume with the claim that eternal truths are tautologies and by closing the last volume with the assertion that what we call thinking is merely a sequence of tautologies.
31 R. Carnap, 'Die alte und die neue Logik,' *Erkenntnis*, vol. 1, 1930, pp. 21-22.
32 L. E. J. Brouwer, 'Mathematik, Wissenschaft und Sprache,' *Monatshefte für Mathematik und Physik*, vol. 36, 1929, pp. 153-64.

Index

Abbe, E., 41-2, 46, 47, 68, 191
abstraction, 41, 97-8, 194
actual, 90
actuality, 118
affirmation, 78, 111
Agassiz, L., 13
aggregative thinking, 109
aggregative view-point, 90, 92-3
analysis, logical, 92-3, 136
analytic philosophy, *see* analytic tradition
analytic tradition, 1-7, 61, 66, 175-6, 178, 185-6
analyticity, 108
anti-empiricism, 102, 117
anti-psychologism, 36, 53-4
a priori concepts, 19, 24, 29
a priori knowledge, 62-3, 103, 175
a priori truth, 13, 19, 38, 43, 101-2
argument, 85
argument-sign, 86
Aristotelian logic, 10, 11, 22, 49, 52, 65-6, 90, 91, 93
Aristotle, 20, 22, 49, 65-6, 81, 83
arithmetic, 171; as analytic, 46-8, 101-2; founded on geometry, 173-5
assertion, 77, 115-16
atomism, 136-7
Austin, J. L., 194-5
Avenarius, R., 37
Ayer, A. J., 5

Bacon, F., 10, 65
Bartlett, J., 191
Bauch, B., 37, 53, 54, 56-7, 59, 173, 179, 192
Bauer, B., 27
Becher, J. J., 50

Becker, O., x
Bedeutung, 130, 144-5, 158
Black, M., xi, 197
Bochenski, I. M., 65
Bolyai, J., 43
Bolzano, B., 40, 43
Boole, G., 66, 91, 180
Boolean algebra, 49, 68-72, 80, 83, 91, 93, 96
Bopp, F., 20
Borges, J. L., 192
Bradley, F., 15, 176
Brentano, F., 35-6
Britzlmayr, W., ix
Brouwer, L. E. J., 3, 175, 184-5
Büchner, L., 17-19, 25, 27

Cantor, G., 8, 43, 98, 129, 162, 191; on sets, 146-7
Carnap, R., xi, 1, 2, 37, 95, 170-1, 178-81, 184, 186
categorial distinction, 143-4
Church, A., 148
class, 97-8, 146, 164
class logic, 80-1
Cohen, J., ix
coloring, 84
complete-incomplete, 93, 138-41
Comte, A., 29-30
concept, cannot be named, 142-3; as function, 56-7, 86, 139, 146; and judgment, 23-4, 30, 49, 55, 90-5, 119-20; as metaphor, 24; and object, 100, 137; and theory, 130-4; traditional interpretation of, 141-2
conditional, 78-80
content, 54-5, 76-7, 84, 86-7

199

INDEX

context principle, 55, 94–5, 124, 133–4, 145, 181, 182, 185–6
correspondence theory of truth, 114
counting, 124–5, 174
Couturat, L., 61, 75, 193
critical positivism, 37
Currie, G., 195–6
Czolbe, H., xi, 17–19, 25, 26–32; on concept and judgment, 30; on traditional logic, 28

Dalgarno, G., 50
Darmstaedter, L., 92, 134, 170
Darwin, C., 18, 34
Dedekind, R., 8, 43, 162
definition, 98; contextual, 127; explicit, 127
denial, 77, 78, 111
Descartes, R., 5, 10, 23, 44, 51, 65, 189
dogmatic positivism, 37
Dilthey, W., 190
Dreyfus, H., x
Drobisch, M. W., 30
Dummett, M., ix, x, 14, 18, 44, 45, 59, 95, 117, 145, 157, 161, 169, 176; on context principle, 133–4; on Frege's historical position, 8–9; on history of analytic tradition, 3–6; on history of semantics, 180–1; on realism in Frege, 105–7

Eggenberger, P., x
empiricism, 6, 12, 14, 55–6, 112
Engels, F., 14, 17
epistemology, 112, 154, 161
equinumerous, 125
equivalence, 150
Erdmann, B., 38
Esperanto, 75
Euclid, 72, 81, 100
Euler, L., 85
existential judgment, 88–90
extension of a concept, 102, 127–8, 146, 147, 155–6, 170, 171
extensional, 197

Feuerbach, L., 14, 17–19, 25, 27
Feyerabend, P. K., ix
Fichte, J. G., 11, 12, 16, 59
Fischer, K., 16–17, 41
formal language, 25, 66, 180, 184–5
formal logic, 12, 49, 52
formal semantics, 1, 3, 6, 95, 180
Foucault, M., 184, 190

Foy, L., 197
Frege, A., 41
function, 85, 99, 130; in *Begriffsschrift*, 139; cannot be named, 142–3; as incomplete, 140–1
functional expression, 86, 140

Gauss, C. F., 42–3, 45, 172
Geach, P. T., xi, 194, 196, 197
general sentence, 87–90
geometry, 44, 171–5; as a formal theory, 1, 171–2; as synthetic *a priori*, 45, 103
Gödel, K., 71, 81–2, 143, 175
grammatical structure, 135
Grassmann, H., 69
Grimm, J. and W., 20
Gruppe, O. F., xi, 1, 19–26, 27, 49, 50, 74, 131; compared with Wittgenstein, 26; on concept and judgment, 23–6; on traditional logic, 22

Hacking, I., 95
Haeckel, E., 34, 179
Hamilton, W., 72
Haym, R., 15
Hegel, G. W. F., 8–9, 11, 12–15, 16, 25, 27, 34, 53, 61, 176
Heidegger, M., 1, 53, 54
Heine, H., 14
Heine, H. E., 162
Helmholtz, F., 188
Helmholtz, H., 19, 29, 162, 188
Herbart, J. F., 40
Herder, J. G., 1, 20, 50
Herschel, J., 30
Hertz, H., 183
Hilbert, D., 43, 46, 71, 171–2
Honderich, T., ix
Hönigswald, R., 173–4
Humboldt, A., 14, 21
Humboldt, W., 20
Hume, D., 18, 91, 105, 178
Husserl, E., 2, 38, 39–41, 54, 162
Huxley, A., 189

idea, 112; as subjective, 117–18
idealism, 8–9, 12–15, 59–60, 107, 176, 182
identity, 98, 122, 126, 137–8, 150
identity statement, 57, 150–4, 159–60
identity thesis, 30–1
if-then, *see* conditional
illumination, 84

INDEX

incomplete, *see* complete–incomplete
induction, 19, 43, 56, 101
innate concept, 24
innate idea, 18
innate truth, 30
intuition, 45–7
intuitionism, 107, 194
Ishiguro, H., ix

Jerusalem, W., 38, 75
Jevons, W. S., 70, 72
judgment, 77, 115–16

Kant, I., 3, 5, 11–13, 19, 30, 32, 33, 36–7, 40, 43, 44, 50, 51, 82, 101–3, 105, 107, 108, 114, 123, 138, 151, 154, 171–4, 176, 179; on the concept of object, 121–2; on existence, 88–9; on geometry, 45; influence on Frege, 58–64; influence on Lotze, 53–7; on unity of judgment, 90–1
Kantian formalism, 61–3
Kircher, A., 50
Kotarbinski, T., 181
Kreisel, G., 106
Kripke, S., 161
Kronecker, L., 8
Kuhn, T., 190

Lachmann, K., 20, 21
Lakatos, I., ix
Lange, F. A., 40, 190
Lange, L., 131–3
Lasswitz, K., 68
law of truth, 113, 158
Leibniz, G. W., 5, 9, 12, 13, 20, 21, 25, 33, 40, 53, 70, 90, 92, 108, 154, 176; on arithmetic, 82; on atomism, 91; on identity statements, 151–3; influence on Frege, 58–64; on logic, 10–11; on logical language, 49–51
level of function, 143
Liebig, J., 14, 15, 18, 19
Liebmann, O., 37
lingua characterica, 71, 191, 192
linguistic philosophy, *see* analytic tradition
Lobatschewskij, N. I., 43
Locke, J., 20
logic, and *calculus*, 71; concept of, 102–4, 108–17; and language, 71; and psychology, 26–7, 112–13; intensional *vs* extensional, 144, 148, 183

logical constituent, 135
logical formalism, 61–3, 109
logical language, 6, 49, 63–4, 116
logical object, 102, 123, 164–5, 173, 175
logical positivism, 37, 178–9, 184
logical structure, 135
logical truth, 108
logicism, 48, 57–8, 110, 173, 175, 183
Lotze, H., xi, 5, 19, 27–8, 29, 40, 41, 49, 59, 63, 76, 103, 104, 122, 130, 154, 173, 176, 192; on Boolean algebra, 72–4; compared with Leibniz, 28; on concept and object, 138; on correspondence theory of truth, 114; critique of Czolbe, 30–2; on formalization, 73–4; on identity, 151–3; on judgment, 77–8; on logic, 52–8; on objectivity, 118–20; as a philosophical traditionalist, 32–3
Löwenheim, L., 180
Ludwig, O., 39
Lullus, R., 50

Mach, E., 1, 175, 179, 183; and logical empiricism, 38–9; and psychologism, 37–9
Marx, K., 14, 17, 27
material conditional, *see* conditional
materialism, 14, 17, 18
mathematical formalism, 18–19, 62, 97, 109, 123
mathematics as a system, 100
Mauthner, F., xi, 1, 19, 39, 175; influence on Wittgenstein, 183–5; on language, 74–6
Mayer, R., 18
meaning, 130, 175
Mill, J. S., 19, 26, 29–30, 154
mode of determination, 153, 196
model-theoretic semantics, *see* formal semantics
Moleschott, J., 15, 17–19, 25, 27
Moore, G. E., 15, 175, 176–7
Müller, J., 14, 18

name, 140, 159–60
nationalism, 33, 192
Natorp, P., 61
natural language, 6, 25, 39, 50, 79, 88, 94, 180, 184–5
naturalism, 6, 14, 15, 17, 18–19, 36, 186
negation, 78
Neo-Kantianism, 19, 36–7, 53, 61, 190

INDEX

Newton, I., 131-2
Nietzsche, F., 39, 190
non-Euclidean geometry, 42, 43, 45, 46, 172
Novalis, 20
number, 97-8; complex, 174-5; as logical object, 123-8; natural, 96, 124-7, 174; real, 134, 196

object, concept of, 102-3, 113, 121-3
objective, 90, 133
objective content, *see* content
objective thought, 112
objectivity, 54-5, 103, 106-7, 111, 117-21; of truth, 115
Oken, L., 13
ontological argument, 88-9
ontological logicism, 110
ontology, 105
Opzoomer, C. W., 30
ordinary language, *see* natural language
Ostwald, W., 75, 193

Papin, D., 70
Parmenides, 58
Pasch, M., 43
Patzig, G., 192
Peano, G., 43, 71, 88, 162, 164, 176
phenomenalism, *see* subjective idealism
phenomenology, 41
philosophy of language, 1-2, 19-21, 39, 185-6
Plato, 20, 49, 58, 118
Platonic theory of ideas, 118-20
Platonism, 105, 106-7, 117, 121; ontological *vs* epistemological, 118-20
polemics, 194
possible world semantics, 180
post-naturalism, 36
predicate logic, 82
predication, 77
proper name, 122
propositional logic, 76-82
psychologism, 2, 8-9, 18-19, 37-8, 97, 104, 117; and language, 39; phenomenalist *vs* physiological, 26-7, 29
Pythagoras, 58

radical empiricism, 18, 29, 38, 74
Ramsey, F. P., 110, 183
Ramus, P., 65
rationalism, 6, 11, 13, 58, 183, 184
real, 90
real number, *see* number

realism, 15, 18, 45, 59, 106-7, 117, 119-20, 161, 176, 182, 195; and anti-realism, 32
reality, 54
reason, 58-9, 120
redundancy theory of truth, 114-16
reference, 130
referring expression, 122, 134, 158-60
Reichlin-Meldegg, A., 16-17
Riemann, B., 43, 46
Russell, B., xi, 2, 15, 43, 61, 71, 81, 163-5, 168, 176-8, 179, 180, 181-2, 183, 184, 186; contradiction, 110, 151, 163-4, 167, 169, 171, 175

saturated-unsaturated, 139, 141
Schelling, Fr. W. J., 11, 12-13, 14, 15, 21, 61
Schering, E., 41, 45, 47
Schlegel, F., 20
Schleiden, M., 15
Schlick, M., 5
Scholz, H., ix, 191
Schopenhauer, A., 183
Schröder, E., 76, 80, 83, 91, 92, 93, 162; on the *Begriffsschrift*, 68-76; on sets, 147-8
scientific naturalism, *see* naturalism
scope, 87
Searle, J., x
semantics, 114, 144, 154, 161
sense-reference, 150, 152-61
sensualism, 28
sentence form, 87
set, 97-8, 146-7
Sigwart, C., 55
Sinn, 130
speculative philosophy and language, 21-3
Spencer, H., 19
spiritualism, 190
Stirner, M., 20
Strauss, D. F., 17, 27
Stroud, B., x, xi, 178
subject-predicate, 83-4
subjective idealism, 31, 104-5
suprasensual, 28, 30, 31-2

Tarski, A., 1, 3, 6, 71, 175, 180-1
tautology, 184, 198
theory of types, 143, 175, 178, 183
Thomae, J., 162
thought, 77, 130; as objective, 105
traditional logic, *see* Aristotelian logic
transcendentalism, 60

202

INDEX

Trede, L. B., 51
Trendelenburg, A., xi, 26, 48–52, 63, 131
truth, concept of, 111–17; indefinable, 113–14
truth-value, 130, 158
Tugendhat, E., 158

use–mention, 152

Vaihinger, H., 190
validity, 120
value-range, 110, 128, 129, 145–6, 149, 156–7, 163–70
variable, 87, 99
Vienna Circle, 178–9

Vogt, K., 17–19, 25, 27
Volapük, 75

Weyerstrass, K. T., 162
Whitehead, A. N., 81
Wilkins, J., 50, 192
wirklich, 118–19, 133, 195
Wittgenstein, L., xi, 1, 2, 3, 19, 26, 44, 116, 144, 163, 175, 179, 181–6, 193, 196
Wolf, A., 20
Wolff, C., 11, 16
Wollheim, R., ix
Wundt, W., 12, 15, 55

Zeller, E., 16

Lightning Source UK Ltd.
Milton Keynes UK
18 May 2010

154375UK00001B/53/P